DETACHED COMMERCIAL ARCHITECTURE

单体商业建筑

(葡)科鲁兹 编著 常文心 译

辽宁科学技术出版社

图书在版编目（CIP）数据

单体商业建筑 /（葡）科鲁兹编著；常文心译. --
沈阳：辽宁科学技术出版社，2014.5
　　ISBN 978-7-5381-8503-4

　　Ⅰ．①单… Ⅱ．①科… ②常… Ⅲ．①商业－服务建
筑－建筑设计 Ⅳ．①TU247

中国版本图书馆CIP数据核字(2014)第034366号

出版发行：辽宁科学技术出版社
　　　　　（地址：沈阳市和平区十一纬路29号　邮编：110003）
印　刷　者：利丰雅高印刷（深圳）有限公司
经　销　者：各地新华书店
幅面尺寸：215mm×285mm
印　　张：12.5
插　　页：4
字　　数：50千字
印　　数：1～1200
出版时间：2014年 5 月第 1 版
印刷时间：2014年 5 月第 1 次印刷
责任编辑：陈慈良　胡嘉思
封面设计：曹　琳
版式设计：曹　琳
责任校对：周　文
书　　号：ISBN 978-7-5381-8503-4
定　　价：208.00元

联系电话：024-23284360
邮购热线：024-23284502
E-mail: lnkjc@126.com
http://www.lnkj.com.cn
本书网址：www.lnkj.cn/uri.sh/8503

DETACHED COMMERCIAL
ARCHITECTURE

单体商业建筑   （葡）科鲁兹 编著   常文心 译

辽宁科学技术出版社

Preface / 前言

Modern commercial space no more just focuses on functional requirements. The functional space, shopping experience and branding strategy are blending, interpenetrating and coexisting constantly. This phenomenon has reflected that people's consumer psychology and concept are changing. The design of a commercial building should consider the combination of natural environment, transportation condition and local features, as well as human behaviours in the environment. The design should consider the issues in the point of human requirements and internal changes. How does the commercial space adapt to the development of commercial activities and combine with urban development and social life and which role should the commercial space play in order to evoke consumers' buying inclination have become urgent issues.

Before 1960s, all the commercial buildings were generally designed as single boxes. The façades were also simply treated with openings, without any unique features. Since 1960s, the rapid growth of commodity economy, the extensive spread of mass culture and the intense impact of pop art all imposed tremendous impact on architects and architectural trends.

With the changes of business operation mode, some commercial brands no longer limit themselves in a same shopping centre with other brands. They select detached build-

现代商业空间已不再是纯粹实现其功能化需求，而其功能空间、购物体验和品牌营销战略也在不断融合、互渗、共生。这种现象也从一个侧面反映了现在人们的消费心理和观念改变。商业建筑的设计既要考虑到与自然环境、交通条件、地域特色的结合，又要考虑环境空间中人的行为，从人的自身需要和内因变化角度看问题。商业空间如何适应商业活动的发展，并与城市开发、公共社会生活相融合，又将以何种角色出现唤起消费者的购买欲望，成为一个亟待解决的问题。

20世纪60年代以前,商业建筑基本都被设计成体形单一的方盒子，外立面设计也是简单的门窗处理，毫无个性。20世纪60年代以后，商品经济的高速发展，大众文化的广泛传播，加上波普艺术的强烈冲击，均对建筑师及建筑思潮产生了巨大的影响。商业建筑设计出现了多元化发展，既有外观简单实用的作品，又有表现奇特，手法创新的作品。

随着商业运作模式的转变，一些商业品牌不再局限于与其他品牌同在一家购物中心里出现，它们选择"独立门户"，有的立足于商业街

ings in commercial blocks, locations with dense clients or busy streets to run their business. The brand client and architect all pay careful attention to the design, in order to provide a comfortable consumer environment which can evoke buying inclination.

Focusing on "detached" and "commercial buildings", this book includes more than 20 excellent projects of different business types from more than 10 countries. These projects interpret basic architectural vocabulary of detached commercial buildings and demonstrate design principles comprehensively. Each project has its unique characteristics: some use unique forms and skillful colour combination to give unique features to the buildings, some emphasise integration with environment, some place the buildings in historical and cultural inheritance, some use new materials and reasonable texture to reinforce the expression, some focus on the use functions, some conform to contemporary requirements of low carbon and energy saving…

The beauty of architecture is a coordinate system with multiple axes. It varies according to different times, local cultures and nationality backgrounds and condenses certain economical and cultural features, as well as efforts and talents of designers. We wish this book will open a gate to understand detached commercial building design for readers, widen their visions and enhance their minds.

区，有的在客户密集处选址，有的建于静谧的街巷。品牌客户和建筑师也在设计上颇为用心，都是为了给公众提供一个舒适宜人的消费环境，唤起消费者的购买欲望。

本书主题意在单体和商业建筑两点，收录了10余个国家的20余个不同商业类型的优秀案例。这些案例从不同角度诠释了单体商业建筑的基本建筑语汇，全景展示了单体商业建筑设计应遵循的一些原则。其中的每个项目都有自己鲜明的特点，它们或以独特的造型和巧妙用色赋予建筑物鲜明的个性，或以建筑与自然环境的融合为重点，或在历史和文化的传承中为建筑物定位，或充分运用新材料和建筑材质的合理使用以强化建筑物的表现力，或以建筑与使用功能的结合为重点考量，或以环保低碳与节能顺应时代的要求……

建筑之美是一个多轴的坐标系。在不同的历史时期，不同地域文化，不同民族背景下都不尽相同，凝结那个时代的经济、文化特征，凝结着设计者的心血和才华。愿这本书的问世能为读者打开一扇了解单体商业建筑设计的大门，达到令读者开拓眼界，精神升华的目的。

CONTENTS 目录

Chapter 1: Commercial Design Integrated with Environment
第一章 与自然环境相互融和的单体商业建筑设计

010 Architecture and Environment
建筑与环境

011 Shopping Environment
购物环境

011 Brand and Design
品牌与设计

012 Shopfront Design
店面设计

Cases Study
案例解析

016 engelbert strauss Workwear Store
engelbert strauss 工作服专卖店

024 Parkpraxis
公园诊所

030 Step Motors Selection
步伐汽车精选店

034 Volvo Experience Centre
贝斯特沃尔沃体验中心

042 Winecenter Winzerhof Dockner
多克纳酒庄

048 PUMA Brand Store, Osaka
大阪PUMA 品牌店

Chapter 2: Architectural Forms and Materials
第二章 建筑造型与材料

056 Architectural Forms
建筑造型

057 Construction Materials
建筑材料

Cases Study
案例解析

064 Armoires Cuisines Action
衣橱烹饪

070 Beans in Kanazawa
金泽豆子书店

076 Blaas General Partnership
BLAAS 博尔扎诺

084 Daikanyama T-Site
代官山茑屋书店

092 HALSUIT
春山西装店

098 Retail Centre Holermuos
霍尔莫斯购物中心

104 J&B Beauty World
J&B美丽世界

112 James Perse Flagship Store
詹姆斯珀思旗舰店

116 Lordelo Pharmacy
洛尔德罗药店

124 Missoni Store
米索尼服装店

128 Placebo Pharmacy
安慰剂药店

Chapter 3: Architecture and Colours
第三章 建筑与色彩

138	Concept 概念	156	Ferrari Factory Store 法拉利工厂店
128	Three Properties of Colours 色彩的三属性	162	Jewellery Showroom for Batukbhai & Sons Batukbhai & Sons 珠宝店
139	Cognition of Colours 色彩的认知	168	Peeps & Company™ 小鸡糖果旗舰店

Chapter 4: Architecture and Sustainability
第四章 建筑与可持续发展

142	Preparations for Architectural Colour Design 建筑色彩设计的前期准备
143	Considerations about Architectural Colour Design 建筑色彩应该注意的几点
143	Colour Categories of Architectural Components 建筑构成要素的色彩分类
143	Reinforcement and Restraint of Architectural Colours 建筑色彩的强化与抑制
144	Construction Materials' Influences on Colour Selection 建筑材料对色彩选择的影响
145	Construction Materials and Colours 建筑材料与色彩
	Cases Study 案例解析
146	Bershka Shibuya 波丝卡涩谷店
152	ETECH Shop Linz ETECH 林茨店

174	Building Envelope 建筑外壳
179	Lighting 照明
180	HVAC Equipment and Systems 空调设备和系统
183	Service Water Heating 生活用水加热
186	Others 其他
	Cases Study 案例解析
188	Climate Protection Supermarket 气候保护超市
194	Municipal Market 市政市场
200	Index 索引

Every building exists in its surroundings. Detached buildings play different roles according to their locations. The secret of harmonious unity between architecture and environment lies in integration. Today's consumers are highly attuned to the importance of branding. It isn't just labels and logos that need to be right, but the entire shopping environment must be pitched perfectly to support and enhance retailers' offerings.

任何一个建筑都处于其所在的环境中。单体建筑要根据其所在的环境担任不同角色。实现建筑与环境和谐统一的奥妙在于融和。如今的消费者十分注重品牌。品牌不仅需要有正确的商标和标识，整个品牌的购物环境都必须对商品销售起到辅助和促进的作用。

COMMERCIAL DESIGN INTEGRATED WITH ENVIRONMENT

与自然环境相互融和的单体商业建筑设计

1.1 Architecture and Environment

Architecture is functional and its spirit is deposited in the entity.

Every building exists in its surroundings. Some are located faraway, without any fundamental facilities. Their design focus on aesthetics and their uniqueness lies in the distinctiveness. However, most buildings coexist with other building groups. We could treat location factor in multiple levels because it concerns whether a building could blend with its surroundings' overall feeling.

Detached buildings play different roles according to their locations. As a leading role, the building should express itself sufficiently and stand out; as a supporting role, the building should learn to coordinate and avoid overriding the leading role's privileges. The distinguishing feature and artistic expression of the building can highlight

1.1 建筑与环境

建筑是具有使用功能的，其精神因素应寄托在实体之中。

任何一个建筑都处于其所在的环境中，有的建筑位置偏远，甚至没有任何基础设施，在设计上主要强调美感，它们的独特之处正是缘于它们的与众不同。而大部分建筑常与其他建筑群共存。我们可以从多层次看待位置这个因素，这关系到一个建筑能否与其所在环境的整体感觉相融和。

单体建筑要根据其所在的环境担任不同角色。做主角时就应该充分的表现自己，能够脱颖而出；做配角时也应该学会配合，不可反宾为主。建筑的个体特色和极具艺术性的表现力可以凸显自己，但并不是每一座建筑都要成为一颗明星，有时，一幢建筑

it. However, not every building should become a star. Sometimes, a building looks ordinary, even dull, by itself; yet its interaction with surrounding buildings achieve a harmonious and appropriate total environment.

The secret of harmonious unity between architecture and environment lies in integration. Coordination is integration and comparison is integration too. The ultimate goal of architects should be "achieving a whole overwhelming the summation of its individual parts".

1.2 Shopping Environment

Brands can be global, but the shopping environment should celebrate the cultural aspects of its location, inspiring shoppers to have fun, take time to enjoy what's on offer and ultimately return time and again. From defining the personality of a major urban regeneration scheme to the finer details that get noticed and give a place its identity, architects need add real value to the projects.

1.3 Brand and Design

Today's consumers are highly attuned to the importance of branding. They are demanding. Brands have to innovate constantly to keep up with the continually changing requirements of the public. Some of the best known brands on the high street to create stores that embody their cultural ethos and values, from ensuring maximum kerb appeal to the right lighting and finishes on the shop floor. It is here, where the consumer meets the products in person, that the brand becomes a reality. The brands should be focused and respondent to the needs of society.

1. The reflective façade of Parkpraxis can reflect the surroundings
2. The clinic is set off by the park green space
3. Wincenter Winzerhof Dockner is built according to the terrain

单独看并不出彩，甚至平淡无奇，但由于与其周边建筑的相互作用，反而会使其在总体环境中显得协调得体、相得益彰。

实现建筑与环境和谐统一的奥妙在于融和。协调是一种结合，对比也是一种结合。建筑师的最终目的应该是实现"整体大于它的各部分的总和"。

1.2 购物环境

品牌可以全球化，但是品牌的购物环境应当凸显它所在地的文化氛围，让消费者享受购物的乐趣，从而再次进行消费。从明确城市复兴规划的特征到注意细节设计并赋予某个地点特定的身份，建筑师需要增加项目的真正价值。

1.3 品牌与设计

如今的消费者十分注重品牌，他们十分苛刻。品牌必须不断改变以迎合公众的需求。一些位于主要商业街上的著名品牌在店铺设计中体现了自身的文化品质和价值观。从路缘石的宽度、合适的照明到地面的装饰，无一不在设计考虑之中。消费者在店铺中亲身接触商品，使得品牌成为现实。品牌必须注重并反映社会的需求。

1. 公园诊所具有反光的外立面，将周围环境反射出来
2. 掩映于公园绿地中的诊所
3. 依地势而建的多克纳酒庄

It isn't just labels and logos that need to be right, but the entire shopping environment must be pitched perfectly to support and enhance retailers' offerings.

1.4 Shopfront Design

Good shopfronts make high-quality contributions to the building as well as to the street scene. Set out below are the broad principles for particular elements of shopfronts.

品牌不仅需要有正确的商标和标识，整个品牌的购物环境都必须对商品销售起到辅助和促进的作用。

1.4 店面设计

优良的店面设计可以创造高品质建筑环境，也能为街景增添色彩。以下是店面设计的一些基本原则。

4. Technical drawing of shopfront design
4. 店面设计技术图

1. 10in HQI 250W multi
   Size: W 275mm* D 130mm* H 150mm
2. T 8mm mirror glass fin.
3. App. gray water paint fin.
4. 20*20 G/V steel plate/app. white emulsion paint fin.
5. T 8mm tempered glass/orange veneer sheet fin.
6. 20*20 G/V steel plate/app. white emulsion paint fin.
7. T 18mm MDF/app. orange emulsion paint fin
8. App. self-leveling fin.

1. 10" 250W卤素灯
   尺寸：宽275mmX直径130mmX高150mm
2. 8mm厚镜面玻璃
3. 灰色水性涂料
4. 20X20 G/V钢板/白色乳胶漆
5. 8mm厚钢化玻璃/橙色层压板
6. 20X20 G/V钢板/白色乳胶漆
7. 18mm厚中密度纤维板/橙色乳胶漆
8. 自动调平层

1.4.1 Doors and Access

The design of the entrance door itself must reflect the design of the other elements which make up the shopfront. Particular attention should be given to the windows such that the bottom panel of the door is of the same height as the stall riser and both door and window frames are of the same material.

Painted timber, two-thirds glazed doors are recommended for shop entrances. Solid unglazed panelled doors are appropriate for access to living accommodation above the shop. Entrance doors and access ramps should be designed to be accessible to disabled people. This means the door should have a clear opening width of at least 750mm and preferably 800mm. If a ramp to the entrance door is needed, it should not exceed a slope of 1:12.

1.4.2 Windows

Large plate-glass shopfronts without any visual support for the upper part of the premises can have a detrimental effect. The window

1.4.1 门和出入通道

入口门的设计必须反映其他店面元素的设计。特别应当注意与门底板同高的窗户，店面底座与门框、窗框应采用同种材料。

建议在店铺入口处采用2/3玻璃镶嵌的彩漆木门。实心门适用于店铺上方住宿区域的入口。入口门和出入坡道的设计应当为残障人士提供便利。这意味着门的净开宽度至少应为750mm，最好可达到800mm。如果门口设有坡道，其坡度不应超过1:12。

1.4.2 橱窗

如果大型平板玻璃店面的上方没有视觉支撑，可能会造成反面效果。窗户应当与建筑的比例相称并略

should reflect the proportions of the building and be slightly recessed within the frame. Timber mullions and glazing bars should be used to break up the window into smaller compartments where appropriate.

1.4.3 Stall Risers

The stall riser provides a visual and structural base for the shopfront and is an essential element of the design. Its height will vary depending on the style adopted, with lower stall risers sometimes taking the form of a deep moulded skirting. The stall riser should have a moulded projecting sill to provide a strong junction with the glass. Stall risers are often timber and panelled but can also be made from glazed tile or marble, but never brick infilled.

1.4.4 Traditional Lettering and Sign Writing

Oversized lettering can give a cluttered and unattractive appearance to the streetscape. The lettering should reflect the proportions of the fascia and the quality and character of the shopfront. Hand-painted or individually fixed lettering (e.g., brass or other metal) will be encouraged. The best option is to use individual letters restricted to the shop name. Clear well-spaced letters are as easy to read as larger oversized letters.

微嵌入窗框。可以用木窗框和玻璃格条来将橱窗划分成小块。

1.4.3 店面底座

店面底座为店面提供了视觉基础和结构基础，是店面设计的基本元素之一。它的高度由设计风格所决定，较低的店面底座有时以深嵌的塑形墙脚线的形式出现。店面底座应有一个突出的窗台来与橱窗玻璃进行连接。店面底座通常为木质镶框式，也可以由釉面砖或大理石制成，但是决不能以砖为填充物。

1.4.4 传统字体和标志文字

过大的字体会让街景显得杂乱而不美观。字体应当与招牌的比例以及店面的品质和特征相称。建议使用手写或独立固定的文字（例如，铜字或其他金属字）。最好使用店名的专属字体。间距适当、简明清晰的文字比过大的字体更易阅读。

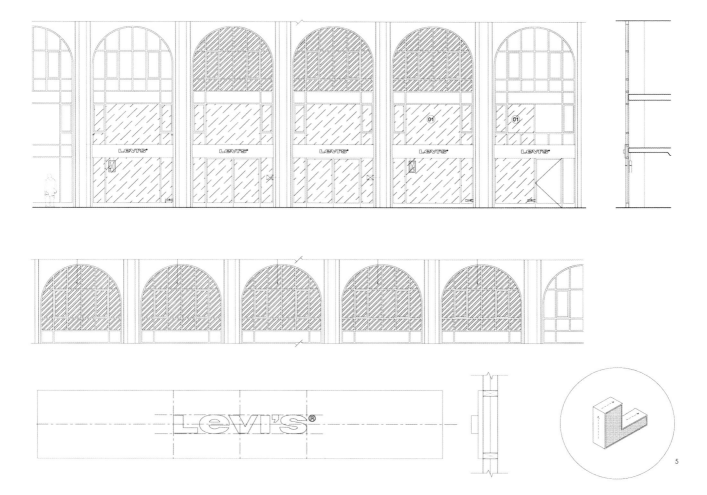

5. Technical drawing of shopfront design　　5. 店面设计技术图

1.4.5 Projecting and Hanging Signs

The authority is likely to approve signs which:
- are in character with the scale of the building.
- are located at fascia level.
- are respectful of the architectural features of the building.
- do not protrude more than 100mm and are not internally illuminated.
- use a style of lettering appropriate to the character of the building.

Highly reflective and brightly coloured plastic signs are inappropriate for historic conservation areas. The use of standard corporate advertising and signs can be damaging in some locations. Stores will be required to show flexibility and consideration to their building and its surroundings.

Plastic and projecting box signs will not be permitted on buildings in conservation areas. They often block the view of other shop signs and are therefore generally discouraged. Hanging signs should not damage architectural features and should be located sensitively at fascia level. It is important that colours harmonise with the detailing and character of the building and surrounding area. Free-standing adverts, such as "A" boards placed on the pavement, are not permitted because they obstruct the public highway.

1.4.6 Blinds and Canopies

The design and material of blinds and canopies is an important element in the character of shopfronts. They protect goods from damage by sunlight and can provide interest and colour in the street scene and shelter for shoppers in bad weather.

Plastic or fixed blinds are not acceptable, and nor are "Dutch" blinds. Blinds and canopies at first floor level and above are rarely satisfactory and will be resisted.
Canvas blinds or canopies of the flat or fan type are usually appropriate but they must be capable of being retracted easily into a recessed area. Existing original canvas blinds and blind boxes should be retained and refurbished. Blinds and canopies should usually be the same width as the fascia but should not cover architectural details.

Lettering may be acceptable where a retractable roller blind obscures the fascia when in use. When included, lettering or symbols should be limited in size. Consent under the Advertisment Regulations may be required.

1.4.7 Lighting

Internally illuminated signs on fascias are often out of place and will be resisted. Shopfronts can be disfigured by a clutter of swan-neck or long-stemmed projecting lamps or crude internally lit fascias. If a fascia is to be lit, it must be done discreetly so as not to detract from the character of the building.

Internal illumination of the fascia and signs is not appropriate in conservation areas. Where lighting is proposed, full details of the fitting, method of fixing and luminance will be required in support of the application. Back lit or halo illumination of fascia signs may be acceptable if well designed. In all cases, external lighting is preferred.

1.4.5 突出和悬挂的标牌

以下几种标牌更易获得管理单位的许可：
- 与建筑比例相称的标牌
- 与招牌在同一水平线上的标牌
- 尊重建筑特色的标牌
- 突出部分不超过100mm、并且不采用内部照明的灯箱标牌
- 字体风格与建筑特色相符的标牌

历史保护区不宜采用高度反光的亮色塑料标牌。标准的企业广告和标牌在某些地点也可能会造成破坏。店铺设计应当考虑到建筑及其周边环境，体现灵活性。

历史保护区的建筑不允许安装塑料和突出的标牌。因为它们会阻挡其他店铺的视野。悬挂的标牌不应影响建筑的特色，并且应与招牌平齐。标牌的色彩应与建筑细节、特色及周边环境和谐一致。禁止使用独立的广告（例如，人行道上的A字板），因为它们会阻塞公路。

1.4.6 百叶窗和天篷

百叶窗和天篷的设计和材料是店面设计的重要因素。它们让商品远离日光损害，同时还能为街景增添趣味和色彩、为购物者遮风挡雨。

不建议使用塑料百叶窗、固定百叶窗和荷兰百叶窗。二楼及二楼以上的百叶窗和天篷的设计大多不尽人意，应谨慎使用。

平面或扇形的帆布百叶窗或天篷通常比较合适，但是它们必须可以轻易伸缩。应当保留原始的帆布百叶窗和百叶窗匣，并对其进行翻新。百叶窗和天篷的宽度应当与招牌宽度相等，不能遮挡建筑的细部设计。

当伸缩式卷帘遮挡了招牌时，可以在卷帘上添加文字。文字或图形的尺寸应符合相关的广告规定。

1.4.7 照明

不建议在招牌上使用内部照明标牌。突出的曲颈或长颈灯以及灯箱招牌都可能破坏店面的整体设计。如有需要，招牌的照明设计应十分谨慎，不能破坏建筑的整体特色。

历史保护区内不适合使用灯箱招牌和标牌。照明设计应全面考虑安装细节、安装方式和照明度。对招牌进行良好的背光照明或光晕照明比较可行。通常来讲，更推荐使用外部照明。

1.4.8 Security

Security should be considered at the design stage. In this way the overall design of the shopfront is enhanced by the unobtrusive inclusion of security elements. By contrast, a well-designed shopfront can be let down by ill-conceived or "add-on" security measures which respect neither the building nor the surrounding area.

All items of security, including burglar alarms and camera surveillance systems, should form an integral part of the design and be located in unobtrusive positions that avoid interference with any architectural detail. Wiring should be internal as far as possible; if external, it should not be visible.

1.4.8 保安措施

店面设计应考虑保安措施。不显眼的保安设施可以提升店面的整体设计。相反，考虑不周或唐突的保安措施会破坏店面的设计，损害建筑乃至周边环境。

所有保安设施（包括报警器、摄像监控系统）应当融于设计之中，设置在不显眼的位置，以避免影响建筑的美观。电线应尽量设在墙内；设在墙外的电线也应遮挡起来。

6, 7. The shopfront design of Puma Brand Store, Osaka
8. engelbert strauss Workwear Store creates a shopping experience that resembles the working environment of customers

6、7. 彪马店店面设计
8. engelbert strauss工作服专卖店营造出的与购买者工作氛围相同的消费体验

	Completion date: 2012	Designer: plajer & franz studio	diephotodesigner.de
016	Location: Bergkirchen, Germany	Photographer: Ken Schluchtmnn /	Site area: 2,400sqm

engelbert strauss Workwear Store

engelbert strauss 工作服专卖店

Key words: The Unity of Brand Theme and Architecture

engelbert strauss positions itself as world-class workwear expert. The store is designed in a solid and heavy industrial form according to the brand's feature, which implies where the products are used. Furthermore, the interior design interprets the concept of "the world of workwear" through simple design, such as tyre sofas, container display shelves, locker display shelves, decorations made of construction materials and tools, etc. All these provide the customers a working-like environment for shopping experience.

主题词：品牌主题与建筑实现的统一

engelbert strauss品牌将自身定位为世界级工作服的供应专家。该项目根据客户的品牌特点将之设计为一个敦实、厚重、极似工业建筑的造型，借建筑之形体体现品牌产品的使用场所。室内设计进一步将"使用场所"诠释，设计基调简洁，配以诸如轮胎沙发、集装箱展示架、更衣柜展示架、施工材料和施工工具的装饰物……使消费者置身于工作环境来实现消费体验。

Design Points:

The store concept for the 2,400m² shopping space convinces with a simple and honest design presenting the world of workwear in an unconventional and surprising way. plajer & franz studio found a playful way to apply references to various trade disciplines into the design concept. By doing so, a customer journey has been created and the store offers its visitors a unique and emotional shopping experience.

The working environment together with its emotional values forms the basis for the store design created by Berlin based plajer & franz studio. The ingenious realisation evokes faith and tangibility for the consumers. At the same time it expresses specialist competence. "Craftsmanship" is real at the engelbert strauss workwear store and not simply a means of decoration as common in lifestyle and fashion stores. It is a "real"

store for "real" craftsmen who are looking for high-quality workwear and appreciate a surprising product display and shopping atmosphere.

Emotional focus points are integrated into the store design with the intention to structure the space and create a customer journey. For example: highlight zones (footwear area, the "strauss island" and the "bird's nest"); theme and focus walls displaying tools, construction materials and other products of workmanship origin; theme-focused ceiling design made out of brushes, tension belts or wrecking balls in the look of mirror balls; and lights created from water-levels or shovels.

Deco elements such as carts transformed into display tables and mirrors, big car tyres used for the presentation of products as well as changing rooms built out of containers further enhance the

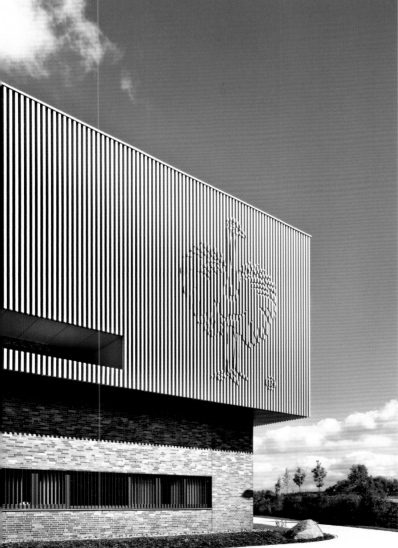

1. Exterior view with surrounding
2. The shopfront design
3. Logo detail

1. 建筑外观及周边环境
2. 店面设计
3. 标识细部

emotional character of the working environment. Just as a side effect they achieve what is most important – smiles on people's faces!

The shopping experience is complete only when you visit the strauss café created from materials and elements of craftsmanship origin. A real highlight for the youngest customers is certainly the bird's nest with an integrated slide!

The new store in Bergkirchen is the first workwear store of this size and dimension and has the role of a prototype for future engelbert strauss workwear stores.

1. South exterior view
2. East exterior view
3. The cashier
4. Underwear area
5. T-shirt area

1. 建筑南侧外观
2. 建筑东侧外观
3. 收银台
4. 内衣区
5. T恤区

CHAPTER 1 **Cases Study** | Detached Commercial Architecture

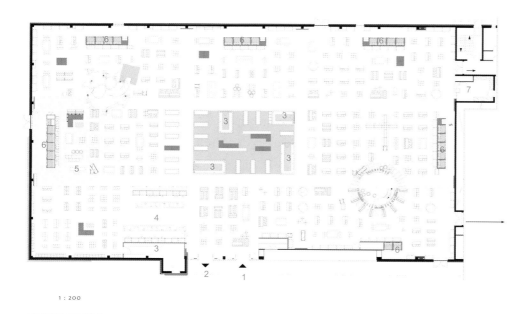

Ground floor plan

1. Entrance
2. Exit
3. Storage
4. Cash desk area
5. Café
6. Fitting rooms
7. Return cash desk

一层平面图

1. 入口
2. 出口
3. 仓库
4. 收银台
5. 咖啡厅
6. 更衣室
7. 退货处

2,400平方米的店铺空间以简洁平实的设计和打破常规的方式呈现出工作服的世界。普拉嘉&弗兰兹工作室以巧妙的方式将商业模式融入到设计概念之中。专卖店将为消费者打造独一无二的购物体验。

店铺设计的基础是工作环境及其情感价值。巧妙的现实化体验唤起了消费者的信心和购买欲。同时，设计也烘托出产品的专业素质和能力。与其他休闲服装店和时装店不同，engelbert strauss 工作服专卖店真正体现了"工艺技术"的精髓。它专为"真正的"工匠提供高品质的工作服，并且拥有令人惊喜的产品展示和购物氛围。

空间结构和顾客体验是店铺设计的两大目标，设计师在其中添加了情绪焦点。例如，重点区（鞋类区、品牌岛和"鸟巢"），展示工具、建筑材料和其他工艺产品的主题墙壁，以刷子、拉力带和模仿铁锤球的镜面球为主题的天花板设计，水准器和铁铲组成的吊灯等。

1. Female area
2. Mirrors
3-5. Showcase detail

1. 女装区
2. 镜子装饰
3~5. 展示细部

CHAPTER 1 **Cases Study** | Detached Commercial Architecture

手推车被改造成了展示台和镜子,汽车轮胎则被用于产品展示,更衣室由集装箱改造而成……这些都进一步营造了工作环境的情绪氛围。最重要的还是这些设计的侧面效果——人们脸上的微笑。

strauss 咖啡厅让整个购物体验变得更加完整,咖啡厅的全部都由工艺技术材料和元素打造而成。最能吸引年轻消费者的当属带有滑梯的鸟巢结构。

这家专卖店是 engelbert strauss 首家同类型的工作服专卖店,它将成为未来 engelbert strauss 店铺设计的蓝本。

1,2. Workwear area
3. Showcase detail
4. Outdoor wear area
5. Showcase detail

1、2. 工作服区
3. 展示细部
4. 户外服区
5. 展示细部

| Completion date: 2012 | Designer: Xarchitecten | Site area: 500sqm |
| Location: Kasten bei Böheimkirchen, Austria | Photographer: Xarchitecten | Construction area: 224sqm |

PARKPRAXIS 公园诊所

Key words: A Combination of Architecture and Natural Environment
This project is a high combination of architecture and natural environment. According to its surrounding natural environment, the designer pays great attention to the form and materials of the architecture. As a result, the project both complete its community function and "hide" itself inside the park. Without violating the continuity of the green spaces, the building loses its independent presence.

主题词：建筑和自然环境的融合
这是一个建筑和自然高度融合的项目。根据建筑所处周边自然环境条件，设计师在建筑形态和建筑材料方面颇费苦心，既满足了该项目对于社区整体功能的补充，又将其完美的"隐匿"在公园之中，丝毫没有破坏绿色空间的延续性，巧妙的消除了建筑的独立存在感。

Design Points:

Task

Dr. Fehrmann decided to build her new practice in the centre of the lower Austrian community of Kasten. The community's centre consists of buildings which surround a green park on three sides. On its fourth side, the park opens up towards agricultural fields and meadows. The premises lie on the edge of the park where its green spaces turn into agriculture. Keeping the valuable urban quality of the continuous green space demands a certain concept in which the building is not to be seen as such, so as not to disturb the geographical connection between the two green spaces.

Concept

The overlapping of the practices' orthogonal, rational room layout with the romantic and free arrangement of the old tree population on the floor plan leads to a geographical penetration between trees and building. The circular tree discs form indentations in the outer shell and round courtyards inside the building. Mirror façades create an additional interconnection between building and park. The park visually doubles in size with its mirror image. The duality between building and nature disappears; the building loses its independent presence and coalesces with its surroundings. This leads to an extension and new interpretation of

the green space while at the same time the service infrastructure becomes denser and positively enlivens the community's centre.

Implementation

The outer façade of the solid single-storey building consists of a curtain wall made of synthetic aluminium composite panels with a reflective stainless steel surface. The building is entered via a circular indentation on the corner. Via the foyer, reception and waiting room, one reaches the circular inner development which runs around the larger courtyard. All of the practice's rooms lead to the completely glazed gallery with its views and exits to the courtyard. From there, two smaller circular rooms ("subcircle distributors") develop further rooms. Each one is exposed according to the concept of overlapping rectangular and round forms via circular shapes.

1. Entrance from outside
2. View from the park
3. Atrium

1. 入口外观
2. 从停车场看建筑
3. 中庭

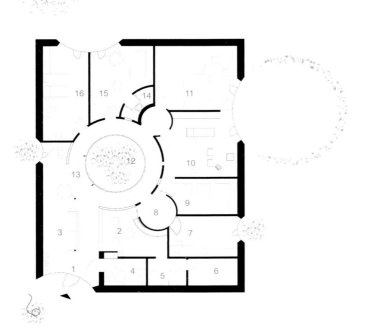
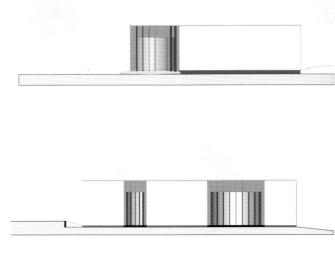

Elevations 立面图

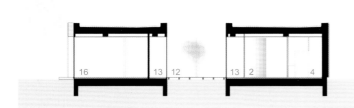

Ground floor plan　　**一层平面图**
1. Foyer　　　　1. 门厅
2. Front Desk　　2. 前台
3. Waiting Room　3. 候诊室
4. Wc　　　　　4. 洗手间
5. Kitchen　　　 5. 厨房
6. Utility Room　 6. 杂物间
7. Laboratory　　7. 实验室
8. Alley　　　　 8. 走道
9. Drugstore　　 9. 药房
10. Ordination 1　10. 诊室1
11. Ordination 2　11. 诊室2
12. Courtyard　　12. 庭院
13. Gallery　　　13. 长廊
14. Bath　　　　14. 浴室
15. Psycho　　　15. 心理治疗室
16. Physio　　　 16. 理疗室

A shady plane tree with three trunks was planted in the big inner courtyard; it being the tree under which Hippocrates had taught his pupils the art of medicine. A small European black pine was planted in the circular concrete entrance area with its swirl finish. The pine's trunk metaphorically forms the stick around which the Aesculapian snake wiggles, the symbol of human medicine. The remaining tree discs with their gravel surface contrast with the surrounding greenness. Pyramid shaped hornbeams were planted in the two smaller circles. An umbrella bamboo hedge was planted to be used as a shield around the big circle in front of the two consulting rooms and the big south-facing circle includes an existing fruit tree as part of the external design.

1. Atrium from inside
2. Waiting room

1. 从室内看中庭
2. 候诊室

CHAPTER 1 **Cases Study** | Detached Commercial Architecture

任务

福尔曼医生决定在奥地利南部的卡斯顿社区中心建造一座新诊所。社区中心的建筑群三面环绕着绿地公园，在另一面，公园朝向农田和草地。项目场地位于公园的边缘、绿地和农田的过渡区。为了保持连续绿色空间那宝贵的城市价值，设计师必须提出一个合适的概念，让建筑既不显眼，又不影响两个绿色空间之间的地理连接。

概念

诊所直角而理性的房间布局不断重叠，与浪漫自由的古树相互配合，在树木和建筑之间形成了渗透。圆形的树木圆盘在建筑外壳上形成了缺口以及建筑内部的圆形庭院。镜面外观在建筑和公园之间形成了额外的互动。公园在镜像中被放大了一倍。建筑和自然的界限消失了；建筑丧失了独立的存在，与周边环境融为一体。这既延伸了绿色空间，又对它进行了新的诠释。与此同时，社区中心的服务性设施变得更密集，使其显得生机勃勃。

实施

结实的单层建筑的外立面由带有复合反光不锈钢表面的合成铝幕墙组成。人们从建筑一角的圆形缩进进入，经过门厅、前台和候诊室，到达圆形的中庭。建筑中心是一个大庭院。所有诊疗室都通往玻璃走廊，从走廊可以欣赏庭院的景色，并直达庭院。走廊里，两间略小的圆形房间可以用作未来使用。两间房通过圆形造型展现了重叠的方形和圆形造型理念。

宽敞的内庭里种植着一棵分为三杈的茂盛悬铃树；希波克拉底（希腊的名医，被称为医药之父）曾在悬铃树下向学生讲述医疗的艺术。圆形的混凝土入口区域种植着一棵小小的欧洲黑松，并配有螺旋形装饰物。松树的树干暗喻着艾斯库拉普的蛇杖——人类医学的象征。其他的树木圆盘都铺设着碎石，与周边的绿地形成了对比。金字塔形的角树被种在两个较小的圆盘之中。两间诊疗室前的圆圈中种植着一片伞竹围栏；而朝南的圆圈中，原有的果树成了外部设计的一部分。

Elevation reflect the surrounding
外立面反射出周围的环境

First floor plan
二层平面图

1. Ordination
2. Waiting room and reception
3. Detail of ordination
4. Psychotherapy room

1. 治疗室
2. 候诊室和前台
3. 治疗室细部设计
4. 心理治疗室

Completion date: 2011
Location: Miki-city Hyogo, Japan
Designer: PROCESS5 DESIGN, Ikuma Yoshizawa, Noriaki Takeda, Tatsuya Horii
Photographer: PROCESS5 DESIGN
Construction area: 165.84sqm

Step Motors Selection

步伐汽车精选店

Key words: Unity of Architecutre and Marketing
The design carefully considered second-hand car customers' feeling, creating a space with imagination, exitement and pleasure. Along with building the second-hand car showroom, a repair shop for after sales services, a sheet metal shop and a car wash were also built. Therefore, customers could enjoy a series service after buying a car, having a sense of security and trust. This is a model of high unity of project and sales target.

主题词：建筑与营销的统一
设计充分考虑了进口二手车购买者的消费心理，营造了一个让人可以产生联想、激动和愉悦感受的空间氛围。于此同时，与之配套的售后修理店、钣金车间和洗车房也随之而建。使购买者能便捷的享受购买二手车以后的一系列服务，增加其购买的安全感与信任感。这是项目与销售目标高度融合的一个典范。

Design Points:

Select car shop where you can choose your lifestyle. This was a project to convert a major car manufacturer's car showroom to a second-hand car showroom specialising in imported cars. Due to the image that a second-hand car dealer sells old things and the image that its products are of less value than those of a showroom that sells new cars, it is a fact that many customers who buy second-hand cars fell like they are making a compromise.

There was a request from the client that they would like customers to buy cars in an easy-going manner according to fashion and they would like them to feel the authentic enjoyment of choosing even a second-hand car.

Therefore, the designers incorporated a sales method similar to that of an apparel select shop that has selected objects for customers with value according to a sense of security and trust and they proposed a select shop that sells cars selected by the second-hand showroom from cars from all over the world.

First of all, the designers constructed the whole showroom with 3 L-shaped walls. The designers displayed photos and sports equipment that remind people of going on holiday in their car on the red brick L-shaped wall which bears a close resemblance to red-brick building and they displayed photos and sports equipment that remind people of cars that not normally visible in everyday life on the white brick wall which bears a close resemblance to a house renovated from a warehouse.

The wooden board L-shaped wall that goes from the approach to inside the showroom is painted with the words "ENJOY CAR LIFE" which was a request from the client and is produced to make a carefree and easy-to-enter space with the combination of the deck, grass and stage that are continued from the exterior. Each different part is a space like a garage, like a living room, like a study, like a lounge or like a restaurant, expressing the lifestyle that each customer dreams of.

1. The display window
2. The café area outside
3. Exterior view

1. 橱窗
2. 露天咖啡区
3. 建筑外观

Through these 3 L-shaped walls, the whole showroom has become a space in which customers can experience the world view of cars and their future lifestyles.

Along with building the second-hand car showroom, a repair shop for after sales services, a sheet metal shop and a car wash were also built. Being aware of a sense of unity with the showroom, the designers proposed a sense of security and a world view that cannot be obtained from a traditional second-hand car showroom from which a customer buys a car and then has no more connection with it by having customers feel the existence of these extra facilities with the continuity of the decor.

The designers aimed to create a shop to express the showiness and excitement of cars in the space overall, to have customers fully feel the enjoyment of choosing a car and to make it possible to select a car according to fashion.

1. The display area
2. The rest area
3. Interior view

1. 展示区
2. 休息区
3. 室内设计

Ground floor plan

1. Show room
2. Salon
3. Reception desk
4. Office
5. Meeting room
6. Kitchen
7. Reception room
8. President's office
9. Lavatory(men)
10. Lavatory(women)
11. Warehouse
12. Machine room
13. Deck terrace
14. Service station
15. Car washing space
16. Visitor parking
17. Display space
18. Employee parking

一层平面图

1. 展览室
2. 沙龙
3. 前台
4. 办公区
5. 会议室
6. 厨房
7. 接待室
8. 董事长办公室
9. 男洗手间
10. 女洗手间
11. 仓库
12. 机房
13. 露台
14. 服务台
15. 洗车房
16. 访客停车位
17. 展示空间
18. 员工停车位

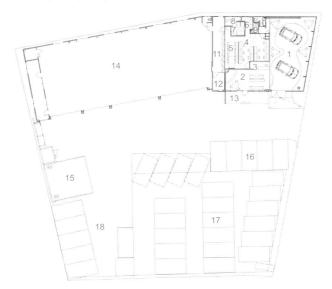

选择一家汽车店就选择了一种生活方式。这个项目将一家主流汽车制造商的汽车展厅改造成了一家专营进口二手车的汽车店。由于人们大多认为二手车销售商卖的都是旧车，其价值低于展览厅卖的新车，许多买二手车的人都觉得自己所选的车不尽人意。

项目的委托人希望他们的消费者能跟随时尚潮流进行轻松的购物，希望他们在选购二手车时的心情愉快。

因此，设计师选择了一种与服装精选店相似的销售方式。精选店的商品让消费者感到安全和信任，因此设计师把销售商从世界各地精选的二手车放置在展览厅内进行销售。

首先，设计师用三面L形墙壁来构造整个展览厅。在L形红砖墙（与红砖建筑的墙壁相似）上，设计师用相片和运动设施来唤起人们在假日驾车出游的热情；在白砖墙（与仓库的墙壁相似）上，设计师则用相片和运动设施展示日常生活中并不常见车型。

根据委托人的要求，通往展厅内部的木质L形墙壁上书写着"享受汽车生活"的字样。木板台、青草和延伸到外面的露天平台营造出轻松愉悦的空间，吸引着人们走进店铺。展厅内的各个部分有的像仓库，有的像客厅，有的像书房，有的像酒吧或餐厅，展示了消费者所幻想的各种生活方式。

这三面L形墙壁让整个展厅成为了消费者体验汽车和未来生活方式的平台。

除了二手车展示厅外，项目还包括售后修理店、钣金车间和洗车房。为了实现展览厅与这些空间的整体性，设计师力求让消费者体验与传统二手车销售的不同。传统二手车销售结束之后，消费者就与店家失去了联系，而该项目的配套设施让消费者感受到了安全感。

设计师旨在打造一个能够展示汽车风采的空间，让消费者在选车过程中保持愉悦的心情，挑选出自己喜欢的时尚爱车。

Completion date: 2008
Location: Best, the Netherlands
Designer: M Sc.Ton Smulders
Photographer: Van den Pauwert
Construction area: 4,000sqm

Volvo Experience Centre
贝斯特沃尔沃体验中心

Key word: An Incorporation of Architecture and Brand
Volvo is a world known automobile brand and this showroom strengthens the brand's value. The hard black exterior design strengthens the brand positioning in safety and steadiness while the partial curves implies humanity and user-friendly design. In interior design, the all-around exhibition provides full-scale experience of every automobile, again confirming the brand's positioning of steadiness and maturity.

主题词：建筑与品牌的结合
沃尔沃是一个国际著名的汽车品牌，这个汽车展厅独特的设计又强化了这个品牌的价值。黑色的坚硬外观设计，强化着这个品牌安全和稳重的品牌定位；局部的弯曲造型，又隐喻着这个品牌的人性化和人情味；在室内设计上，对于汽车的展示也进行了全方位的设计，使消费者能够全面的体验每一台汽车的使用感受，再次印证了品牌稳重、成熟的定位。

Design Points:

On the highway A2, exit at Best-South near the city of Eindhoven, stands a large solid-looking building dressed in black. The grand signing leaves nothing to clarity, you've reached Volvo-dealer the Beemd. In a horizontal slid alongside the highway a few Volvo models stand as a showcase ogling passing motorists. It is a glimp of the Volvo experience that is waiting for you inside.

Approaching the building via the main route you are lead around the curved shape ending in a parking place underneath the round edge of the building. Here the experience begins and the building gets more and more open revealing its interior.

From here a lazy staircase leads you into the dense volume that gives room to a number of functions not immediately expected in any ordinary car showroom. A lounge, meeting facilities and a kitchen studio contribute to the special input of this showroom.

In the centre of the building, the main stage has a

central position. Here one has an overview over the inner levels of the building, with the various showrooms for new and used cars. Even some oldtimers can be seen. Above the stage the building opens itself to the air, daylight flows in. On the top level is a small patio.

Inside the building the cars get all the attention they deserve. The attention is not distracted by looks outwards, but the closed walls offer a quiet background, incidentally supplemented with projected images that reflect the atmosphere of a specific model, speed and dynamics, or the unique selling points such as the enormous amount of storage space and the attention for safety and security of the driver.

A car showroom should not look like an aquarium, where you can examine cars from the outside. It belongs to an environment where the car can show all its aspects, so that after seeing, feeling and smelling the cars, people can make a well-thought-out decision, just as the experience of the car that Volvo embodies.

1. Cambered exterior in night
2. Square exterior
3. Cambered exterior

1. 曲面外观夜景
2. 方形外观
3. 曲面外观

1. The glass display area
2. The ramp
3. The window

1. 玻璃橱窗展示
2. 坡道
3. 窗户

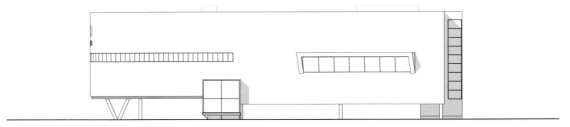

South elevation 南立面图

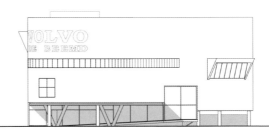

West elevation 西立面图

在荷兰 A2 高速公路的贝斯特南出口，靠近埃因霍温市，有一座高大敦厚的黑色建筑。建筑商巨大的标识再清楚不过，这是一家沃尔沃汽车销售店。沿着高速公路的水平滑块上展示着几辆沃尔沃模型，向过往的车辆进行展示。这只是沃尔沃体验的一角。

通过主车道接近建筑，顾客会被引领环绕弧形建筑一圈，最后到达建筑下方的停车场。建筑的内部体验在这里又更接近了一步。

人们从楼梯走进功能繁多的紧凑空间。你会发现这与普通的汽车展览厅不尽相同。展廊厅特别设有酒廊、会客室和小餐厅。

建筑中心的主展台占据了主要位置。人们可以在这里尽览建筑的各个楼层——各色展厅中展示着大量新车和老车。展览厅中甚至还有一些古董车。建筑的主展台向上开放，引入了太阳光。建筑顶层有一个小露台。

建筑内部处处以汽车为主角。封闭的墙壁提供了干净的背景，不会让人受到外界的影响。墙壁上时不时地会出现一些投影图像，反映汽车特殊的模型、速度、动力或是其他独特的卖点，如巨大的仓储空间、卓越的安全性能等。

汽车展示厅不应像一座水族馆，仅仅让人们从外部进行观察。它应当提供一个展示汽车方方面面的环境，让顾客在观看、感觉、体验汽车之后，做出正确的购买决定，正如沃尔沃汽车展示厅一样。

East elevation 东立面图

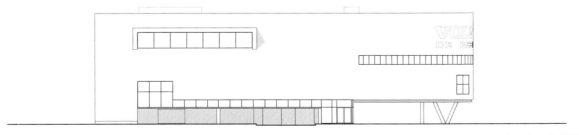

North elevation 北立面图

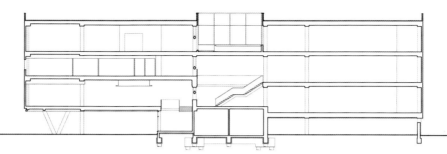
Section 剖面图

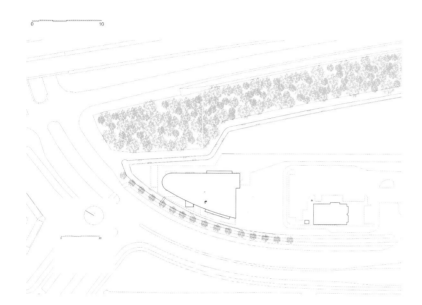
Site plan 总平面图

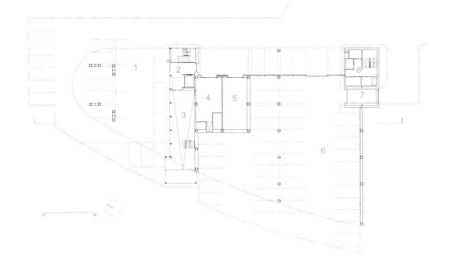

Ground floor plan

1. Public parking
2. Main entrance
3. Ramp
4. Service
5. Car-wash
6. Closed parking
7. Car elevator
8. installation

一层平面图

1. 公共停车场
2. 主入口
3. 坡道
4. 服务区
5. 洗车房
6. 封闭式停车场
7. 汽车电梯
8. 固定设备

1. The glass display area over the water
2. The car showed
3. The café

1. 水上玻璃橱窗展示
2. 汽车展示
3. 咖啡厅

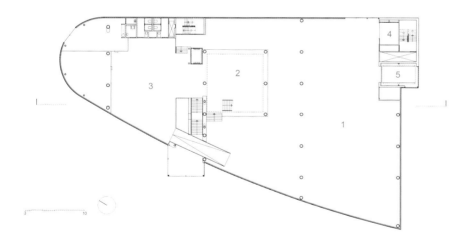

1st floor plan

1. Showroom new cars
2. Podium
3. Inspiration
4. Meeting room
5. Car elevator

二层平面图

1. 新车展览室
2. 展台
3. 灵感区
4. 会议室
5. 汽车电梯

1. Rest area outside
2. Stair
3. Rest area outside

1. 露天休息区
2. 楼梯
3. 露天休息区

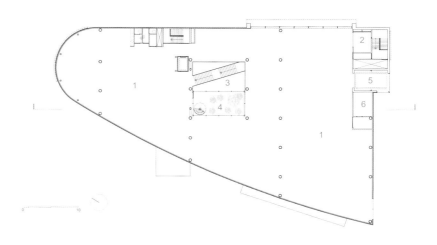

2nd floor plan

1. Showroom used cars
2. Office
3. Server
4. Archive
5. Meeting room
6. Void
7. Car elevator

三层平面图

1. 旧车展览室
2. 办公区
3. 服务器机房
4. 档案室
5. 会议室
6. 上空
7. 汽车电梯

3rd floor plan

1. Showroom used cars
2. Meeting room
3. Void
4. Patio
5. Car elevator
6. Installation

四层平面图

1. 旧车展览室
2. 会议室
3. 上空
4. 天井
5. 汽车电梯
6. 固定设备

Completion date: 2011
Location: Höbenbach, Austria
Designer: Lukas Göbl – Office for Explicit Architecture
Photographer: Boris Steiner
Construction area: 500sqm

Winecentre Winzerh of Dockner
多克纳酒庄

Key words: A Perfect Combination with the Surrounding Topography
Set in the idyllic Krems valley landscape, the building's modern and changeful image is impressive. The surrounding village is organised in various geometries and the building meshes dynamically into the topography. The undulating and changeful shapes interpret the surroundings perfectly. The glass fibre reinforced concrete panels of the façade are reminiscent of loess soil, merging the building into the site.

主题词：依势而建完美融合
项目坐落在田园诗般的克雷姆斯谷景观处，建筑造型现代、多变，给人留下很深刻的印象。周边的村庄环境特点明显——地形呈几何状分布，该建筑依地势而建，起伏错落和变化的造型很好的诠释了周边环境，玻璃纤维钢筋混凝土面板的外观让人联想记忆中的黄土，自然而然的融入环境之中。

Design Point:

Set in the idyllic Krems valley landscape, the contemporary architecture of the Dockner Winery store and tasting centre makes a very impressive statement.

Nestled perfectly into the immediate surrounding topography, the wine centre's contemporary design is an interpretation of the various geometries typical to the surrounding village. The building draws its overall shape from the contours of the land, meshing dynamically into the spatial context of Lower Austria's highly traditional landscape.

The glass fiber reinforced concrete panels of the façade are reminiscent of loess soil. Exposed concrete interior walls draw upon the various levels of the winery operation's terrain, manifesting a strong sense of place. The shape of the building embodies the vintner's visionary approach. The focus of attention here is wine making, with an emphasis on presentation and marketing. Also important is the versatility of the building, which functions as a showroom, wine sales centre, tasting room, and event venue.

The spatial order of the two-storey building into three functional areas is clear and efficient. The areas for wine tasting, sales, and presentation are located on the ground floor. The central wine bar is immediately noticed upon entering, and is flanked by display cases filled with the latest winery specials. Opposite the bar, seating areas following the line of the building create the wine tasting area. An etched glass room divider separates the integrated office from the tasting area, yet still allows for transparency from all vantage points. Two floor-to-ceiling windows enhance the welcoming atmosphere of the tasting room.

The rear showroom and storage area has been stylistically minimalised, devoted entirely to the presentation of the wine. The bottles of wine stored directly in the showroom give it the charismatic atmosphere of a wine cellar. A generously

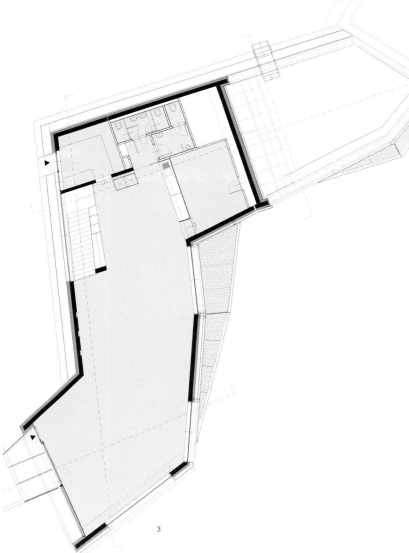

1. The view from southeast
2. Entrance from outside
3. First floor plan

1. 建筑东南侧外观
2. 入口外观
3. 二层平面图

proportioned show window on the back wall provides a view of the new distillery in the old section of the winery.

The upstairs rooms are available for different events – seminars, presentations, lectures, and various festivities. A floor-to-ceiling glass structure with an integrated sliding door offers a beautiful panorama of the Benedictine monastery and the surrounding wine country. A garden is accessible from the multi-purpose room and can be used seasonally as an extension of the interior.

The high standards that characterize the Dockner family's wine production are also evident in the interior of the building, whose entire design was carefully selected and custom-made for the building. The color coordination between the interior and the exterior of the building restates the nature of the winery: birch veneer in Bordeaux red, hand-planed bog oak floor, moss-colored leather, and anthracite lighting elements. The luxurious surfaces of these precious materials is illustrative of the careful detailing and thorough design of this building.

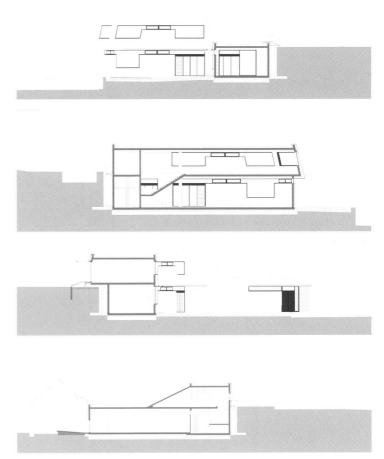

Sections　剖面图

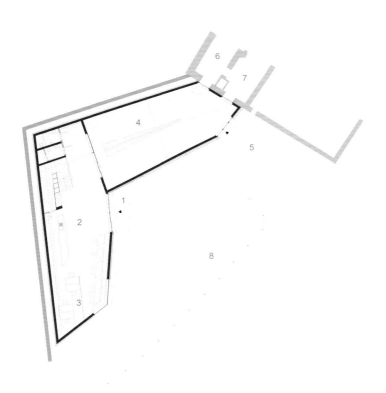

Ground floor plan

1. Main entrance
2. Tasting room
3. Office
4. Showroom
5. Delivery entrance
6. Corridor
7. Distillery
8. Parking area

一层平面图

1. 主入口
2. 品酒室
3. 办公区
4. 展览室
5. 送货入口
6. 走廊
7. 酿酒室
8. 停车区

1. Tasting area and office
2. Wine bar
3. Main entrance

1. 品酒区和办公区
2. 红酒吧
3. 主入口

Site plan 总平面图

CHAPTER 1 **Cases Study** | Detached Commercial Architecture

当代建筑多克纳酒庄和品酒中心坐落在田园诗般的克雷姆斯谷景观，给人留下非常深刻的印象。

葡萄酒中心完美地融入周围的地形，极具现代感的设计是对周围如变化的几何图形般的村庄的完美诠释。建筑的整体形状沿着土地的轮廓延伸，与奥地利传统景观的空间要素相协调。玻璃纤维钢筋混凝土面板的外观让人想起记忆中的黄土。裸露的混凝土内墙利用各层酒厂操作地带表现出强烈的场所感。建筑物的形状体现着酿酒师的远见卓识。这里的重点是葡萄酒酿造，强调的是演示和营销。同样重要的是建筑的多功能性，它的功能包括陈列室、葡萄酒销售中心、品酒室和活动场地等。

这个两层楼的建筑，空间秩序是明确有效的，共分为三个功能区。品酒室、销售区和演示区位于底层。进入后立刻就会注意到中央的酒吧，两侧是最新的精品酒展示柜。沿着对面的酒吧休息区线就到了品酒区。蚀刻玻璃的房间隔板将综合办公区和品酒区隔开，但仍允许从其他有利位置看起来要有透明度。两个落地窗户为品酒室增添了温馨的氛围。

后面的陈列室和存储区被适当淡化了，而完全专注于葡萄酒的演示。葡萄酒瓶直接存储在陈列室里以营造酒窖的氛围。后墙的窗子比例很宽，可以看到酿酒厂的新蒸馏间。

楼上的房间可以举办不同的活动，如研讨会、演讲、讲座以及各种庆祝活动。一个从地板到天花板的玻璃结构带有一个综合性推拉门，可以看到修道院的全貌和整个葡萄酒之乡。花园可以通过多功能室进入，也可以根据季节性作为室内部分的延伸。

多克纳家族的葡萄酒生产以"高标准"著称，这一点在建筑内部也表现得很明显，整个设计都处理的很仔细，并且严格按照客户的要求进行。建筑物内部和外部颜色之间的协调重述了酒庄的性质，其中包括：波尔多红的桦木板、人工刨面沼泽橡木地板，青苔色的皮革和无烟煤照明设备。这些珍贵的材料使设计尽显奢华，这座大楼的每处细节都经过了精心的设计。

1. Stairs to first floor
2. Stairs to first floor
3. Wine consumption space
4. Showroom

1. 通往二楼的楼梯
2. 通往二楼的楼梯
3. 红酒消费区
4. 展示厅

CHAPTER 1 **Cases Study** | Detached Commercial Architecture

Completion date: 2013
Location: Osaka, Japan
Designer: plajer & franz studio
Photographer: die photodesigner.de
Site area: 579sqm

PUMA Brand Store, Osaka

大阪 PUMA 品牌店

Key words: Brand Image and Design Vocabulary
The architects use design vocabulary corresponsive to the business concept for brand positioning and identification in both exterior and interior design. Set off by red meshed metal baffle, natural and artificial light changes and penetrates day and night. The interior is well structured with flexible use of various materials, presenting a smoothing and changing space. Both the exterior and interior design express PUMA brand's concept: happy sports and humorous lifestyle.

主题词：品牌形象与设计语汇
无论是外观和内部空间设计，建筑师都用与商业理念相呼应的设计语汇将品牌定位与识别充分体现。在红色轻质网状金属挡板的掩映下，自然光线和灯光在昼夜交替间变幻、穿梭；室内空间流动多变，层次分明，配以多种材料的灵活使用。展示了PUMA这一品牌一种能体现运动快乐和诙谐精神的生活方式。

Design Points:

plajer & franz studio under the direction of Ales Kernjak (head of global store concepts, Puma retail ag) developed a retail concept that is simple and flexible while integrating local references and offering Puma's clients a joyful shopping experience.

Thereby, the new brand store is more than just a shopping place. It is a social and cultural meeting point. It is a space for events and happenings of various kinds. The lower two floors of the three-storey Puma building have been designated for the 600m² shopping space, while the upper level – an open roof top, surrounded just by a light façade construction – creates an open space for performances and sport events.

An impressive, cone-shaped staircase in the centre of the store with a red brand-wall in its back, is the eye-catcher making a strong brand statement. It draws attention not only to its design but above all to the footwear tribune at the base of it. The "stage" continues as a footwear catwalk stretching across the entire ground floor, which is also home to the sport

lifestyle product line and black label. Although the design of both areas is simple and functional with references to Japanese architecture (e.g., Origami seating elements in the sport-lifestyle area or the use of materials such as wood and concrete), both speak different visual languages and are clearly distinct from each other. With a very light and flexible design, the use of materials such as black steel, plywood panels combined with re-used gym flooring elements and the mixture of matte and high gloss surfaces, the black label area carries a visible sports heritage.

The first floor is dedicated to the performance area. Here, the nature of performance is visible and perceptible throughout. The design for this area is even more flexible and functional, using more technical materials such as stainless steel or

1. The shopfront
2. The glass display window
3. The corner exterior view

1. 店面
2. 玻璃橱窗
3. 建筑转角

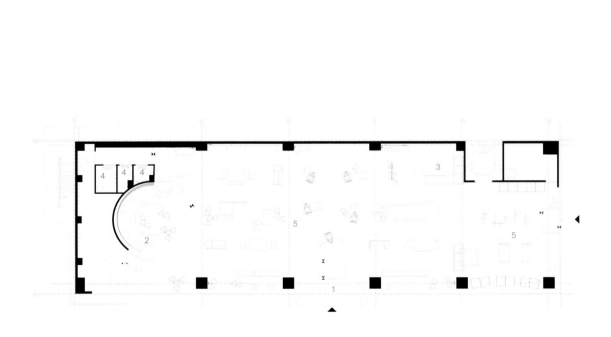

Ground floor plan
1. Entrance
2. Staircase
3. Cashier
4. Changing room
5. Marketplace

一层平面图
1. 入口
2. 楼梯
3. 收银台
4. 更衣室
5. 卖场

perforated metal in silver grey. The flexible approach is further reflected in the design of the display tables, which are covered with convertible materials. A catwalk area runs across the entire performance floor with a stage in the centre of it and leading towards the footwear focus wall.

Generally, shopping at the new premium store is above all an experience and in line with Puma's philosophy an active interaction with the brand. Whether creating their own customised sneakers at the Puma factory, engaging with Puma's sustainable actions summarised in the sustainability journey or enjoying the online experience via ipads, clients shall be surprised, entertained and given opportunities to identify with the brand.

1. The ground floor interior view from the staircase
2. Overlook of the staircase
3. A red brand-wall with the cone-shaped staircase

1. 从楼梯俯瞰一层内景
2. 俯瞰楼梯
3. 红色商标墙和锥形楼梯

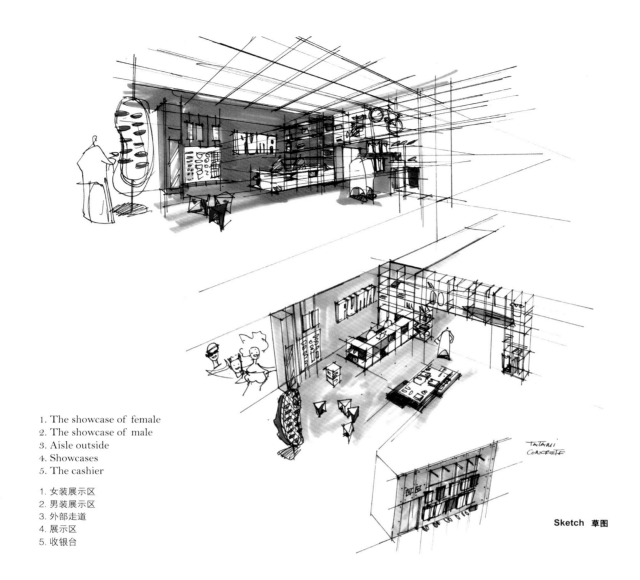

1. The showcase of female
2. The showcase of male
3. Aisle outside
4. Showcases
5. The cashier

1. 女装展示区
2. 男装展示区
3. 外部走道
4. 展示区
5. 收银台

Sketch 草图

在PUMA公司全球品牌概念总监阿尔斯·科恩贾克的指导下，普拉嘉 & 弗兰兹工作室打造了一个简单而灵活的零售概念。设计综合了本地元素，为消费者提供了愉快的购物体验。

因此，新建的品牌店不仅是一个购物场所，更是社交和文化聚集地，举办各种各样的活动。商店大楼的下面两层是 600 平方米的购物空间，而顶层的开放式屋顶则被轻质立面结构围成了一个用于表演和运动的开放空间。

轻质网状金属挡板组成的立面让日光得以渗透到店内。夜晚，室内的灯光透射出来，照亮了街道，也隐约显示出店内的场景。室内设计将运动、表演和品牌形象结合起来，旨在强化品牌形象，提升产品展示效果并突出品牌主打的鞋类产品。

店铺中央的锥形楼梯以红色商标墙为背景，既吸引眼球又表达了强烈的品牌氛围。人们不仅会注意到楼梯设计，更不会忽略下方的鞋类展示。鞋类展示台贯穿了整个一层空间，旁边还展示着运动生活产品和PUMA黑标系列产品。尽管两个区域的设计参考了日式建筑，简单实用（例如，运动生活区的折纸座椅元素、木材和水泥的使用等），二者的视觉效果却截然不同，泾渭分明。黑标产品区的运动元素十分明显，设计轻巧灵活，采用了黑钢、胶合板和再利用的健身房地板，并以磨砂表面和高光表面形成了对比。

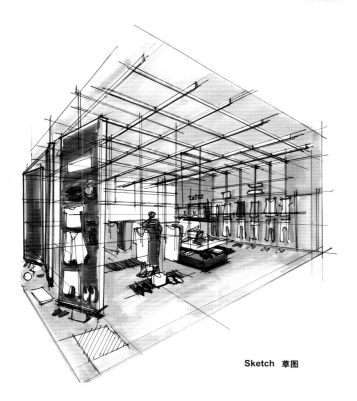

Sketch 草图

二楼主要用作表演区，到处都流露出表演的氛围。这一区域的设计更加灵活使用，采用了不锈钢、银灰色网眼金属板等技术材料。展示台的设计运用了可变形材料，十分灵活。T台区贯穿整个表演楼层，一直延伸到鞋类展示墙。T台的中间设有一个舞台。

总而言之，在这家新开的精品店内购物是一种愉快的经历，能够与PUMA品牌进行互动。无论是在PUMA工厂里定制运动鞋、参与PUMA可持续之旅或是通过上网享受在线体验，消费者都能在惊喜、愉快的过程中了解PUMA品牌。

From creative design to completion, an architect will expend large amount of energy in a project. The continuous development in new technologies and materials provide on one hand more choices for architects, on the other hand more challenges. An appealing and attractive appearance of the building is an important incentive for consuming.

一个项目从创意设计到完成,建筑师倾注了很多的心血。新技术、新材料的不断进步,一方面丰富了建筑师的选择,另一方面也让选择更富挑战。一个建筑从外观上能够吸引大众并为他们所喜爱,是引发消费的重要条件。

ARCHITECTURAL FORMS AND MATERIALS

建筑造型与材料

2.1 Architectural Forms

2.1.1 Definition

Architectural form is the form that the architect designs in accordance with the building's functions, which should take practical internal structures, reasonable layout and external aesthetics into consideration.

2.1.2 Objectives of Architectural Forms

A good architectural design is the unity of content and form and is the harmonious unity of human being, architecture and nature. Considering architectural function as priority is a common principle of modern architectural language. Objectives of architects' creation include practicability, aesthetics, safety and economy. If a building with appealing form and impressive appearance cannot meet the users' requirements or fail to function, it is still to be considered a design failure.

2.1.3 Unique Forms for Individual Commercial Buildings

Unique structural forms will lead into unique architectural forms. Various forms of spatial organisation can create novel effects on architec-

2.1 建筑造型

2.1.1 建筑造型的概念

建筑造型是建筑师根据使用功能对建筑物本身的内部结构实用、布局合理、外部结构美观而设计出来的形态。

2.1.2 建筑造型的宗旨

好的建筑设计是内容和形式的统一，是人、建筑、自然三者的和谐统一。将建筑功能作为首要考虑因素是建筑学现代语言的普遍原则。实用、美观、安全、经济是建筑师创作时应遵循的宗旨。建筑形态引人入胜，造型摄人心魄，而内部功能不能满足使用者的要求或者不能发挥作用，将成为设计中的败笔。

2.1.3 独特的单体商业建筑造型

独特的结构形式会创造出独特的建筑造型，各种形

1. Beans in Kanazawa is designed with beans concept

1. 形如豆子的金泽豆子书店

tural forms. For the design and final implementation of architectural forms, different architects will possess different understandings and analyses on the same project. The individual commercial buildings included in this book are mainly single-brand commercial buildings with small scales. Provided the functional requirements are achieved, the effects of architectural forms play an important role. To some extent, with unique forms, these buildings can attract more consumers, thus realising the owners' requirements and goals.

2.2 Construction Materials

2.2.1 Definition and Classification of Construction Materials

Construction material is the general term that describes the materials

式的空间组合反映到建筑造型上会产生新颖的效果。对建筑造型的设计及最终实施，不同的建筑师会对相同的项目有着不同的理解和剖析。本书中的单体商业建筑多指小体量的单品牌商业性作品，在满足其使用功能的前提下，建筑造型的作用显得尤为突出，从某种程度来讲，这类建筑的独特造型将会招揽更多顾客的光顾，从而实现业主的需求和目的。

2.2 建筑材料

2.2.1 建筑材料概念和分类

建筑材料是土建工程中所用材料（水泥、砂、石、木材、

2. The construction of Armoires Cuisines Action uses various materials
3. J&B Beauty World has a unique exterior

2. 运用了多种材料的衣橱烹饪橱柜店
3. 造型独特的 J&B 美丽世界

(e.g., cement, sand, stone, wood, metal, pitch, synthetic resin, plastics, etc.) used in civil engineering. Construction materials can be classified in three categories: structural materials, decorative materials and some special materials. Common structural materials for buildings include wood, bamboo, stone, cement, concrete, metal, bricks, ceramics, glass, engineering plastics, composite materials, etc.

2.2.2 Principles for Selection of Construction Materials

Function or style? This is an inevitable choice for an architect in design process. Throughout the design process, the architect should always take design, function and comfort level as standards for important choices and considerations.

Good exterior wall materials should be tested for insulation, ventilation, fire prevention, corrosion resistance and wind resistance. Therefore, exterior walls with tested materials can last a long service life. Besides preventing walls from influences of severe weather such as wind, rain and snow, good construction materials should insulate the building from direct sunshine and noises, and maintain a stable level of interior temperature and humidity, creating a comfortable interior atmosphere.

2.2.3 A Table for Construction Materials

金属、沥青、合成树脂、塑料等）的总称。建筑材料可分为结构材料、装饰材料和某些专用材料。常用于建筑的结构材料，包括木材、竹材、石材、水泥、混凝土、金属、砖瓦、陶瓷、玻璃、工程塑料、复合材料等。

2.2.2 选择建筑材料的宗旨

要实现功能，还是要设计风格？建筑师在设计过程中不断的在做这种选择。在整个设计过程中，建筑师要时刻将设计、功能实现和舒适度作为重要的选择和考量标准。

优良的外墙材料应经过保温、通风防火测验、抗腐蚀、抗风载测验等，在这样的材料包裹下的建筑物外墙才会历久弥新。优良的建筑材料除了要保护墙体免受风、雨、雪等恶劣天气的影响，还要使建筑物免受日晒、噪音的困扰，维持一个稳定的室内温度和湿度水平，从而营造一个舒适的室内环境。

2.2.3 建筑材料简表

4. The unique façade of Placebo Pharmacy is composed of different textures and skins
5. Composition diagram of Placebo Pharmacy's façade

4. 不同肌理、表皮组成独特的外立面——安慰剂药店
5. 安慰剂药店外立面构成图

Construction detail drawing
1. Suspended façade made of bamboo rods
2. Steel structure
3. Curved steel plates

结构细部图
1. 竹竿制成的悬垂立面
2. 钢结构
3. 曲面钢板

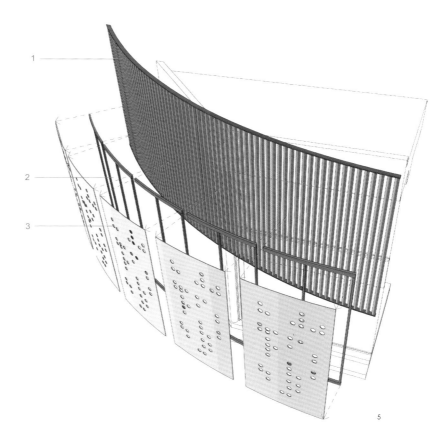

Introduction of Common Exterior Wall Materials 常用外墙材料简介

Position 使用部位	Classification 分类	Advantages 优势	Disadvantages 不足
外墙 Exterior Wall	外墙涂料 Exterior Wall Coating	较为经济,整体感强,装饰性良好、施工简便、工期短、工效高、维修方便,首次投入成本低。即使起皮及脱落也没有伤人的危险,便于更新、替换,满足不同时期建筑的不同要求,进行维护后可以提升建筑形象。在涂料里添加防水剂可以一次施工就解决防水问题。外墙涂料在环保、生产、装饰、施工、对建筑的保护、安全性、经济开支等方面都占有明显的优势,如外墙漆不存在光污染、热污染的问题,老化后的漆膜能够被环境所降解 。另外,外墙涂料不存在对土壤、煤炭等不可再生资源的消耗。可以选择任何一种颜色。 Economical, a strong sense of wholeness, good decoration, easy and short construction, high efficiency, convenient maintenance, low initial cost, harmless even when peeling or falling off, easy renovation and replacement. It can meet the requirements of building in different periods and can improve the building image after maintenance. When added with waterproof agent, the coating can solve waterproof issues once and for all. Exterior wall coating possesses obvious advantages in environment protection, production, decoration, construction, building protection, safety and economy. Without any issues of light pollution and heat pollution, the paint film can be degraded by the environment. In addition, exterior wall coating doesn't consume non-renewable sources such as soil and coal. The colour selection is flexible.	耐候性、自洁性、耐久性、耐霉变性、耐沾污性一般或较差,维护周期短。质感较差,容易被污染、变色、起皮、开裂。同时,寿命较短,不适合在环境污染大的区域使用。 Mediocre or poor weather resistance, slef-cleaning ability, durability, mildewing resistance and contamination resistance, and short maintenance interval. The coating's texture feels not good and it is easy for contamination, discolouration, peeling and cracking. Besides, with short service life, exterior wall coating is unfit in areas with severe environment pollution.
	清水混凝土 Fair-faced Concrete	"绚烂之极归于平淡",最高级的审美就是自然。清水混凝土极具装饰效果,所以又称装饰混凝土。在混凝土施工中经脱模之后所呈现的自然纹理传出一种其他人工建筑材料无法模仿的天然质朴与厚实,借四季的变化呈现不同的气质,成为最具东方文化色彩的建筑材料,带来返璞归真的视觉享受。它浇筑的是高质量的混凝土,而且在拆除浇筑模板后,不再作任何外部抹灰等工程。它不同于普通混凝土,表面非常光滑,棱角分明,无任何外墙装饰,只是在表面涂一层或两层透明的保护剂,显得十分天然、庄重。 Simplicity is Beauty. The highest level of beauty is the natural. Fair-faced concrete is decorative, so it is also known as decorative concrete. The natural texture generated in the process of demoulding expresses inimitable natural simplicity and solidness. Fair-faced concrete presents different qualities in different seasons. It is known as a unique oriental construction material, which brings pure and natural visual enjoyment. It is made of high-quality pouring concrete; after demoulding, it is not treated with any external plastering. Different from normal concrete, fair-faced concrete is smooth and has distinct angles. Without any decorations but one or two layers of transparent protective agent, it appears natural and decent.	科技含量和对施工工人的要求都很高。施工要求高,一般需做保护层以提高耐候性、耐久性、自洁性,并因此增加成本。 Rigid requirements for both technology and construction workers. Generally, fair-faced concrete need a protective layer to improve its weather resistance, durability and self-cleaning ability, which causes higher cost.
	清水砖砌体 Plain Brick Masonry	观感自然,耐候性佳、耐久性好、热工性能好,维护周期长。 Natural appearance, good weather resistance, durability and thermal performance, long maintenance interval.	施工要求高。 Rigid requirements for construction.
	陶瓷面砖 Ceramic Facing Tiles	价格便宜,施工简便,耐候性好,自洁性好,耐久性好。外墙面砖有施釉和不施釉之分。坚固耐用,具备很好的耐久性和质感,色彩鲜艳而具有丰富的装饰效果,易清洗、防火、抗水、耐磨、耐腐蚀和维护费用低。通体砖由岩石碎屑经过高压压制而成,表面抛光后坚硬度可与石材相比,吸水率更低。 Cheap, convenient construction, good weather resistance, self-cleaning ability and durability. Exterior ceramic tiles can be divided into glazed tiles and unglazed tiles. Ceramic tiles are sturdy and durable, with good durability and texture. The colourful tiles are highly decorative. They are easy for cleaning, fireproofing, waterproofing, wear resisting, corrosion resisting and have a low maintenance cost. Full body tiles are made from rock fragments after high-pressure moulding. The flintiness of the polished surfaces can be compared to that of stone. The water absorption of ceramics is lower.	首次投入成本较高,粘贴要求较高,施工难度大,施工质量难以控制,施工技术不过硬容易造成脱落伤人。同时,必须另外采用防水材料解决防水问题。从环保的角度讲,清洗过程中用酸会对大气造成污染。用面砖的外墙一旦发生渗水问题,较难找到渗水的位置,这会给以后的维护带来很多麻烦。 High original cost, rigid requirements for sticking, difficult for construction and uncontrollable construction quality. Waterproof agent must be used to solve leakage problems. In ecological view, the acid used for cleaning will pollute the air. Once the exterior wall has a penetration problem, it is hard to find the exact leaking location, which brings troubles for future maintenance.
	干挂法面砖及空心面砖 Dry-hang Facing Bricks and Hollow Bricks	视觉效果好,耐候性佳、耐久性好、热工性能好,施工方便,无湿作业。 Good visual effect, weather resistance, durability and thermal performance, convenient construction and non-wet construction.	价格高。 High price.

Position 使用部位	Classification 分类	Advantages 优势	Disadvantages 不足
外墙 Exterior Wall	石材幕墙 Stone Curtain Wall	视觉效果好，耐候性佳、耐久性好、自洁性好。可以有效地避免传统湿贴工艺出现的板材空鼓、开裂、脱落等现象，明显提高了建筑物的安全性和耐久性。可以完全避免传统湿贴工艺板面出现的泛白、变色等现象，有利于保持幕墙清洁美观。在一定程度上改善了施工人员的劳动条件，减轻了劳动强度，也有助于加快工程进度。 Good visual effect, weather resistance, durability and self-cleaning ability. It can avoid hollowing, cracks and falling off issues in traditional wet sticking process effectively and enhance the building's safety and durability.	造价高，对瓷板厚度要求较高，一般瓷板不小于15mm厚。开槽时容易崩边，损耗大。需要安装钢龙骨，加大建筑物承重，且增加成本。保温要单独安装，保温安装与幕墙安装交叉施工，安全隐患大。板材的受力方式为点承重，抗震性能差，安全系数低。 High price and rigid requirements for the thickness of the plate, at least 15mm. Large losses in the process of cutting. The required steel joist will increase the architectural load and construction cost. The installation of insulation and stone curtain wall should be separated, which will cause potential safety hazard. The load carrying uses point bearing, which has poor seismic performance and low safety level.
	金属板材（铝、铜、不锈钢、钛、锌板等） Sheet Metal (e.g., aluminium, copper, stainless steel, titanium and zinc sheets)	视觉效果好，耐候性佳、耐久性好、自洁性好。到目前为止，金属幕墙中的铝板幕墙一直在金属幕墙中占主导地位。轻量化的材质，减少了建筑的负荷，为高层建筑提供了良好的选择条件；防水、防污、防腐蚀性能优良，保证了建筑外表面持久常新；加工、运输、安装施工等都比较容易实施，为其广泛使用提供强有力的支持；色彩的多样性及可以组合加工成不同的外观形状，拓展了建筑师的设计空间；较高的性能价格比，易于维护，使用寿命长，符合业主的要求。 Good visual effect, weather resistance, durability and self-cleaning ability. Up to now, aluminium curtain wall plays a dominant role in metal curtain walls. Its light weight reduces the building's load and provides good conditions for high-rises. The good waterproofing, antifouling and anticorrosive properties ensure an enduring external surface for the building. It is convenient for processing, transportation and installation, which provides compelling support for its extensive uses. The various colours and shapes extend architects' design spaces. With a good cost performance, metal curtain wall is easy for maintenance and has a long service life, which fully meets the owners' requirements.	材料的质量不尽相同，防水密封要求严。板背面需设加强筋，以增加板面的强度和刚度。因复合型面板材料的折边只保留了正面板材厚度，厚度变薄，强度降低，所以折边须有可靠的补强措施。造价比较高。 With various qualities of the materials, the requirements for waterproof sealing are rigid. Reinforcing ribs should be added to the back of the sheets to increase the strength and stiffness. The folds of composite sheet materials only retain the thickness for front sheet and the lack of thickness causes lower strength. Therefore, the folds need reliable reinforcement. In addition, it is high-priced.
	防腐木材 Anti-corrosive Wood	视觉效果亲切自然、环保、安全；防腐、防霉、防蛀、防白蚁侵袭。提高木材稳定性，防腐木对户外木制结构的保护更为主要；防腐木易于涂料及着色，根据设计要求，能达到美轮美奂的效果；防腐木接触潮湿土壤或亲水效果尤为显著，满足户外各种气候环境中使用15~50年以上不变。 Warm and natural visual effect, environment-friendly, safe; corrosion prevention, mildew proof and temite proof. With higher stability, anti-corrosive wood is more important for outdoor wood structure. Easy for painting and colouring, anti-corrosive wood can achieve an excellent effect according to the design requirements. It possesses a great hydrophily and can serve for more than 15-50 years in various outdoor environments.	价格高，耐候性、耐久性、自洁性一般。由于加压过程中浸注药剂，木材会略有发绿的化学药剂颜色，并有颜色不均、色差大、无光泽，随着化学药剂的流失，还容易变灰、黑。且一般未采取二次干燥，木材本身的含水率高，因此易出现变形、开裂的问题。防腐木材受环境、温度湿度及太阳紫外线照射，一般使用一年后需要做维护或者涂刷油漆等，还要定期的去维护，费财力又费人力。而防腐木是按标准长度批量生产的，实际应用时存在裁减，自然损耗较多。 High price and mediocre weather resistance, durability and self-cleaning ability. The reagent used in pressurisation will cause undesirable green colour, irregular colour, colour differences and lack of gloss. With the running of the reagent, the wood will become gray and black. Without redrying process, the wood with high moisture content will deform and crack. With influences from surroundings, temperature, humidity and ultraviolet irradiation, generally anti-corrosive wood need maintenance or painting once a year. In addition, anti-corrosive wood is manufactured in standard length and has to be cut in practical use, causing large amount of normal loss.

Introduction of Common Exterior Wall Materials 常用外墙材料简介

Position 使用部位	Classification 分类	Advantages 优势	Disadvantages 不足
外墙 Exterior Wall	室外用人造胶合板 Exterior Man-made Plywood	显示出天然材料自然亲切的视觉效果，与防腐木相比，板材单元面积大，自洁性好。 Warm and natural visual effect. Compared to anti-corrosive wood, plywood has a larger unit area and better self-cleaning ability.	价格高，耐候性、耐久性一般。 High price and mediocre weather resistance and durability.
	玻璃幕墙 Glazed Curtain Wall	耐候性佳、耐久性好、自洁性好。玻璃幕墙是当代的一种新型墙体，它赋予建筑的最大特点是将建筑美学、建筑功能、建筑节能和建筑结构等因素有机地统一起来。建筑物从不同角度呈现出不同的色调，随阳光、月色、灯光的变化给人以动态的美，并具有轻巧美观、不易污染、节约能源等优点。幕墙外层玻璃的里侧涂有彩色的金属镀膜，从外观上看整片外墙犹如一面镜子，将天空和周围环境的景色映入其中，光线变化时，影像色彩斑斓、变化无穷。在光线的反射下，室内不受强光照射，视觉柔和。它可吸收红外线，减少进入室内的太阳辐射，降低室内温度。 Good weather resistance, durability and self-cleaning ability. As a new type of wall, glazed curtain wall can unite architectural aesthetics, functions, energy-saving and structure organically. The building presents various colours in different directions. With the changes of sun light, moon light and lighting, it expresses a dynamical beauty. Glazed curtain wall is light, artistic, contamination resistant and energy-saving. The inner side of the external glass is coated in coloured metal film, creating a mirror effect. With the change of light, the mirrored images present multiple colours and countless variations. Through the reflection, the interior is protected from glare. Glazed curtain wall can absorb infrared rays to reduce interior solar radiation and lower interior temperature.	造价高、墙体热工性能差、能耗较大，产生的反射眩光形成光污染。 High price, poor thermal performance and high energy consumption. The reflected glare can cause light pollution.
	U形玻璃 U-profile Glass	价格中高，视觉效果特殊，耐候性、耐久性、自洁性好。U形玻璃是一种高效的、符合现代建筑设计要求及工业化要求的建筑型材；U形玻璃具有透光不透明的特性，拥有自承重的优势；U形玻璃规格多样，可以根据需要采用不同的组合形式和排列方法；具有独特的光影效果和典雅的颜色质感；是一种新型环保的建筑材料；大面积采用U形玻璃的建筑极富文化气息，平静的立面中蕴含着变化，简洁的造型中透着意境，室内明亮而不耀眼，既能隔音又可保温。U形玻璃建筑模糊了墙与窗的定义，消融了线与面的界限。U形玻璃隔热效果好，特别适用于较热和较冷的地区。 Medium price, special visual effect and good weather resistance, durability and self-cleaning ability. U-profile glass is an efficient construction material that meets both the requirements from modern architecture design and industrialisation. It is light-transmitting and non-transparent. It is self-supporting. The diversified specifications can be united and arranged in accordance with practical requirements. It possesses unique light effect, elegant texture and excellent colours. It is also environemt-friendly. Extensive use of U-profile glass will create a cultural atmosphere which expresses changes in still surface and deep meanings in simple form. The interior is bright without glare and well insulated. U-profile glass blurs the definitions of wall and window, and melts the boundaries between line and plane. With good thermal insulation, U-profile glass is especially applicable to areas with hot or cold climate.	玻璃是典型的脆性材料，其破坏具有突然性，对于一般玻璃来说，当发生火灾时，易受到明火或者辐射热而破裂，从而导致火焰蔓延，且碎片通常呈尖角和锐边，所以在人群活动频繁的场所必须考虑其安全性。该材料易碎，施工过程中损失较大，一次性投入大。 Glass is a typical fragile material with sudden destructive potential. In case of fire, common glass tends to crack because of open fire and radiating heat, which will cause flame spreading. Furthermore, the fragments are sharp, which is unsafe for places with crowded population. The fragileness will also cause losses in instalation and increase original cost.
	玻璃砖 Glass Brick	价格中高，视觉效果特殊，耐候性、耐久性、自洁性好。 Medium price, special visual effect, good weather resistance, durability and self-cleaning ability.	施工要求高，墙体热工性能一般。 Rigid requirements for construction and mediocre thermal performance.

Position 使用部位	Classification 分类	Advantages 优势	Disadvantages 不足
外墙 Exterior Wall	真石漆 Stone Paint	真石漆是近几十年出现的新型建筑外立面材料。真石漆属于碎石颗粒漆膜，自重较轻，危险程度大大降低；漆膜不厚，并不增加墙体自重负担而改变其力学状况；翻新容易，费用较低。真石漆有底漆、线条漆、主漆和罩面漆，既有装饰效果，又保护了墙体不碳化、不渗水；真石漆喷涂随意，能造出各种石材不能轻易制造的造型及符合任何基层状况的墙面；真石漆造价经济，是石材望尘莫及的，大大减低了业主负担；真石漆色泽一致，天然碎石色彩可保百年以上。 Stone paint is a new material for exterior façade of the building. Made from stone particles, stone paint is light and relatively safe. The paint film isn't thick so it doesn't add the wall's weight nor change its mechanical condition. The renovation is easy and low-cost. Stone paint includes base paint, line paint, main paint and finish-coat paint, It will decorate the wall and protect it from carbonisation and penetration. Stone paint can produce various forms that real stone cannot easily produce and it fits walls with various surface conditions. The economical price of paint stone is incomparable by real stone, which will reduce the owner's load significantly. The colour of stone paint can last for more than 100 years.	脱皮，漆膜含有自重较大的碎石颗粒，膜厚比普通乳胶漆厚得多，经常出现真石漆漆膜与墙体表面粘附不牢而有局部掉下来的现象；遇雨水泛白。传统真石漆被雨水浸泡2小时后，发生乳皂化效应，太阳出来之后，皂化的白色乳液封闭在漆膜乳液内，导致漆膜部分发白，影响了仿石的庄重效果；太阳暴晒发黄，影响仿石的装饰效果；开裂。传统真石漆质地坚硬，硬则不韧，随着墙体基层的龟裂，真石漆的部分漆膜也连带龟裂，这对墙体装饰效果而言是极其不利的；易污染，因为真石漆漆膜表面往往粗糙，这样很容易积灰、吸尘、粘灰等污染。 Because the paint film contains large stone parcels, it is much thicker than ordinary emulsion paint. Therefore, the paint is easy to peel once the film is poorly adhered to the wall. The paint will whiten in the rain. After 2 or more hours of rain-soak, traditional stone paint will be saponified. When the sun comes out, the saponified liquid is sealed in the paint film, causing whitening of the film, which influences the decent stone-like effect. Long exposure to sun will cause the paint to be yellow, influencing the decorative effect. As the paint film is hard and inflexible, part of the film will crack with the cracking of the wall surface, which is also harmful to the decorative effect. The surface of stone paint film is always rough, which is easily contaminated by dust.
屋面 Roof	混凝土平屋面（其上做防水处理）Concrete Plat Roof (With waterproof layer)	价格便宜，施工简便，其上一般需设防水层，维护周期短；为满足视觉效果，提高热工性能，可做绿化屋面，地砖铺地等。 Low price, convenient construction and short maintenance interval. Waterproof layer should be installed. To achieve better visual effect and improve thermal performance, the roof can be treated with green roof, tile paving, etc.	视觉效果差。 Poor visual effect.
	瓦屋面（烧结瓦、油毡瓦、彩钢板瓦）Tiled Roof (Fritted tile, Asphalt tile, Colour steel plate tile)	造价相对低廉，施工简便，大部分瓦屋面自然亲切视觉效果好，维护周期一般较长，热工性能好。采用瓦屋面与其他屋面相比较，屋面造型多样，色泽艳丽；加快了屋面施工进度，几乎不受气温的影响；屋面的防水性能大大改善，便于维修检查；与普通平屋面相比较，经济成本大约约节省25%左右；瓦屋面美观大方，取材容易，经济实用；由于坡度大，雨水排流较快，屋面不积水，屋面防水效果较好。 Relative low price and convenient construction. Most tiled roof has a warm and natural visual effect, long maintenance interval and good thermal performance. Compared with other types of roofs, tiled roofs enjoy diversified forms and bright colours. Tiled roofs accelerate the construction progress and are hardly influenced by temperature. The waterproof ability is improved so it is convenient for maintenance. Compared to an ordinary flat roof, a tiled roof will save about 25% cost. The tiled roof possesses an elegance appearance and is economical and practical. With steep slopes, the tiles roof has a better drainage, so the roof has little surface water and a better waterproof effect.	施工时间久，工程造价高，施工单位不按正常施工方法施工或局部施工，会造成使用寿命低、结合处漏水和雨水倒流。 Long construction time and high engineering cost. The wrong implementation by the constructors will cause short service life, leakage in joints and rainwater backflow.
	玻璃、PC板顶棚 Glass and PC Board Roof	常用于共享大厅等大空间部位以引进自然光，观感好。 Commonly used in large areas such as sharing halls to introduce natural light. It has a good visual effect.	价格高，热工性能差，需注意灰尘积聚。 High price and poor thermal performance. Dust deposition should be noticed.
	金属板材（铝、铜、钛板等）Sheet Metal (e.g., aluminium, copper and titanium sheets, etc.)	视觉效果好，防水效果好，耐候性、耐久性、自洁性好，常采用轻型屋盖结构有利于减小建筑自重，维护周期相对长。 Good visual effect, waterproof effect, weather resistance, durability and self-cleaning ability. Light-weighted roof structures are commonly used to reduce the structural weight. The maintenance interval is relatively long.	价格高。 High price.

Completion date: **2010**
Location: **Boucherville, Canada**
Designer: **Jean Verville architecte & Maryse Crôteau designer**
Photographer: **Martin Tremblay**
Construction area: **929sqm**

Armoires Cuisines Action
衣橱烹饪

Key words: A Mixture of Materials to Create a New Style
The project transforms an abandoned building into a showroom for a kitchen cabinet brand. The designer chooses wood, glass, light aluminum panels as materials to form a minimal, bright and generous style, making the store a unique view along the highway.

主题词：材料混搭营造建筑新格调
该项目要将一个已被遗弃多年的建筑改建为橱柜品牌展示店，设计师选择使用木材、玻璃、轻型建筑铝合金板进行交替拼接、交错，组合新店简洁、明朗、清新、大方的格调，使其成为高速公路边一道独特的风景。

Design Points:

This project consists of renovating over 90% of a 10,000 square feet building in Boucherville, Canada. The building, abandoned for about 5 years, is in poor condition. The decision is to demolish it while saving the foundation and the ground floor structure. This implied to keep the implantation position, restore parking spaces and exterior landscaping in a sustainable and economic way.

The programme for this new building is to receive the showroom of Armoires Cuisines Action, a family company manufacturing kitchen cabinets. The goal is to have a structure with a dynamic and contemporary look in order to become a signal in this dense trade sector. Half of the surface is the kitchen cabinet's showroom and the second half is occupied by sales consultant's offices, administration and training areas. Clients aspire to have a building with sufficient transparency so that one can feel the action inside and even see kitchens from the highway.

The reutilization of a part of the old building provides an eco-sensitive and economic approach.

The existing foundations offer a basement with 16 feet high ceiling, which allows an imposing one-floor structure with a total height of 26 feet. In order to optimise the space in the basement some courts are hollowed on either side of the building and plenty of windows are installed. This provides a more dynamic space in the basement, which can now be used it as showroom.

Wooden volumes that seem to penetrate the building bring off the transparency effect. This amplifies the relationship between interior and exterior and accentuates the dynamism of the proposal. The repetitive module, 18 cubic feet by 18 feet, animates the facades as much as the

1. Wood blocks and silver clad form a strong composition that provides a powerful image to Armoires Cuisine Action new showroom.
2. The curtain wall system contributes to abundant natural light, maximises visibleness from the highway, and optimise solar gain. The wooden cube on the roof hides mechanical equipment and increases the building's volume.
3. Courts are hollowed on either side of the building to optimise the use of the basement with its 16 feet high space.

1. 木板结构和银色包层的组合为新展厅提供了强有力的形象。
2. 幕墙系统保证了充足的自然采光，增加了店铺在公路上的辨识度，并且优化了太阳热增量。屋顶的木结构将机械系统隐藏起来并且增加了建筑的体量。
3. 建筑两侧各有一个庭院，优化了16英尺（约4.9米）高的地下室空间。

interiors. The cube on top of the roof conceals the building's mechanical equipment and gives a new volumetric dimension by increasing the total height of the structure.

The abundance of natural light illuminates the entire interior space while the plan brings about the fluidity. All offer constant drilled on the exterior that indisputably contribute to the feeling of openness. The windows, carefully positioned to provide the highest visibility, are proportioned to maximise solar gain.

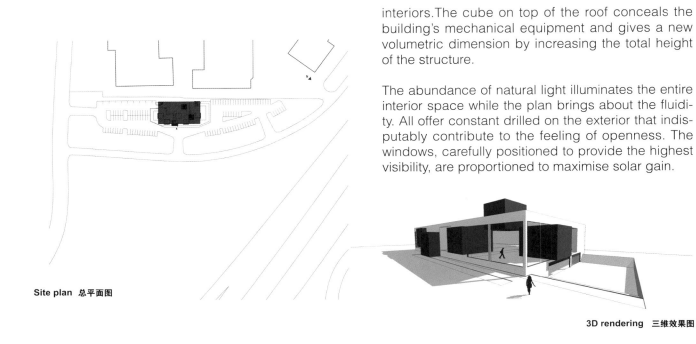

Site plan 总平面图

3D rendering 三维效果图

Elevations 立面图

本项目是对加拿大布谢维尔的一个1万平方英尺大楼进行改造，改造面积超过90%。大楼本身位置并不好，并且被搁置了5年之久。最终我们决定在保留地基和一层结构的基础上进行拆除改造。这意味着要采用可持续发展和节约的方式来保持植入位置，重建停车场和外墙景观。

大楼将要展示加拿大橱柜品牌"衣橱烹饪"。我们的目标是设计一个充满活力和现代感的外观结构，成为这个密集的贸易部门的一个标志。大楼中该品牌的展示区域占一半面积，另一半则是销售顾问的办公室，行政区域和培训区。客户渴望整个空间足够透明以便感受到里面正进行着的活动，甚至是在过道里就可以看清厨房内部。

1,2. The building is like a set of blocks, simple and bold, with a repetitive pattern that animates the façades. The graphic duo tone is obtained with volumes of dark torrefied lime wood and light aluminum architectural panels.
3. Windows provide maximum solar gain and bring transparency. With the hollowed court, they allow for the basement to be used as showroom space.
4. Wood reminds of the production material for kitchen cabinets and brings a new contemporary look to Armoires Cuisine Action new building.

1、2. 建筑像积木搭建而成，简洁而大胆，重复的组合模式令外立面富有生气。深色烘烤椴木和浅色的铝制建筑嵌板在色彩上形成了鲜明的对比。
3. 窗户保证了日光照射的最大化，并且让空间显得通透。它们与凹陷庭院一起，保证了地下室能够作为展示空间。
4. 木材暗示着厨房橱柜的生产材料，为衣橱烹饪橱柜店的新建筑带来了富有现代感的外观。

大楼保留部分的再利用采用的是生态而经济的方式。

现有的地基提供的地下室带有16英尺高的天花板，使得整个一层结构高达26英尺，气势恢宏。为了优化空间，我们在地下室两侧挖出了几块草地，并且安装了很多窗户，这使得地下室成了一个充满活力的空间，可以用来展示商品。

木制材料似乎为建筑带来了透明效应，放大了内部和外部之间的关系，并突显"活力"这一主题。大量18立方英尺的模块使得建筑外立面像室内装饰一样充满生气。屋顶上的立方体隐藏了建筑物的机械设备，增加了建筑结构的高度，并提供了一个新的体积空间。

1,2. Wooden volumes penetrate the building and amplify the relationship between interior and exterior, while industrial materials are used for their expressive qualities.
3. Textures and materials blend in light colours and natural finishes.
4. The metallic staircase structure unfolds down to the lower level.
5. Large windows bring natural light to a 16 feet high basement which is now being used as showroom space.

1、2. 木结构穿透了建筑，增强了室内外空间的联系，而工业材料的使用则更富表现力。
3. 纹理与材料混合了淡薄的色彩和天然装饰。
4. 金属楼梯结构一直延伸至底层。
5. 高大的窗户给作为展示空间的地下室带来了自然采光。

Floor plans and sections
1. Entrance
2. Showcase
3. Bathroom
4. Meeting room

平面图和剖面图
1. 入口
2. 展示厅
3. 洗手间
4. 会议室

充足的自然光照亮整个室内空间，而我们的设计理念是要给整个建筑带来流动性。所有的光照毫无保留地落在建筑外墙，使得这个大楼有一种开阔感。窗户的位置被细心地安排在能提供最高可见度的地方，以便最大限度地提高太阳能增益。

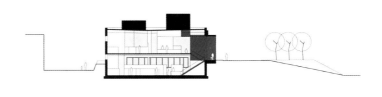

Completion date: **2007**
Location: **Kanazawa, Japan**
Designer: **Keiichiro Sako**
Construction area: **4,250sqm**
Site area: **16,650sqm**
Award: **2008 Japan GOOD DESIGN Award**

Beans in Kanazawa
金泽豆子书店

Key words: Beans with Profound Meanings
Located along an arterial road of Kanazawa, the project is the biggest roadside bookstore in Japan. The beans forms of the building is smooth, soothing and interesting. With the function of a bookstore, the beans can imply seeds, which absorb nutrition from books to grow up. Beans are also introduced into the interior design: pierced patterns outside the cashiers, carpet and book section signs all uses bean patterns. "Reading stands" with different sizes form different windows shapes, which decorate three big "bean" floors, creating a lively atmosphere.

主题词：寓意深刻的"豆子"主题造型
该项目位于金泽市中心的主干道边，是日本最大型的路边书店，整体呈"豆子"造型，建筑外轮廓流畅、舒缓，饶有情趣。借由书店的功能定位，"豆子"造型可引申为种子之意，汲取书籍的养分而茁壮成长。"豆子"之形也引入了室内，收银台外围的镂空图案、地毯设计、书籍分区指示牌都取了此形。大小不同的"站立式阅读台"形成不同的窗户形态，装点着这个三层的大"豆子"，整体感觉更加跳跃、活泼。

Design Points:

Recently, bookstores tend to upgrade their sizes. Bookstore with a whole floor plate near the station or attached with a spacious parking lot along the road in the suburbs is common. Beans in Kanazawa are one of them. With 800,000 books for sale, it can be called th e biggest roadside bookstore in Japan.

The bookstore enjoys an interesting shape. Three masses of beans are stacked together, drawing soothing curves. Without anydead angles inside, the interior spaces are flowing. Along with the walls, bookshelves extend to the roof, creating an endless intellectual atmosphere. For the convenience of selecting and reading books, available spaces are equipped with "reading stands". Reading stands are actually formed through cutting a part of wall shelves at one-metre height. In the meantime, the cut-off part become windows on the wall. At night, passers-by could see people walking inside the bookstore. The design is inspired by "magazine corner" of convenience store, creating a lively atmosphere around the building.

Standing reading and leisure reading are both paid enough attention. Sofas and benches are everywhere. The designer designs some big armrests for the bookstore to place some books. Besides, children colouring book areas on the top floor are covered with colourful carpet, where children could both play and read. The bookshelves are placed radically around the central counter. People could see the whole floor in the centre without any blind spots, which could both avoid book stealing and children's safety. Sofas are placed around the counter, so customers on the sofa could also take care of the children and books. Theme of beans are everywhere: beans can be found on both sides of the counters on the first two floors, floor tiles in the garden, individual poster boards and logos.

1. Southwest exterior view in night
2. Southeast exterior view in night
3. Reading area

1. 建筑西南侧外观夜景
2. 建筑东南侧外观夜景
3. 阅览区

近年来书店趋于大型化。例如车站附近占地一层的大型书店或者郊外路边拥有宽阔停车场的书店成为主流。"金泽豆子"就是藏书80万册的堪称日本最大型的路边书店。

书店的外型富有情趣,把豆子形状的体块叠加3层,并且勾画了舒缓的曲线,内部没有任何死角,形成了流动的空间。书架延墙面设置,直达屋顶,营造出没有尽头的知性氛围。同时为了便于取阅书籍,只要空间允许就会设置"站立阅读台"。阅读台其实是将墙面书架距地面1米高处切割掉一部分,做成平台,切断部分同时充当了墙面的窗口,到了夜晚可以透过窗口看到书店内人头攒动的景象。这种设计灵感来源于便利店的"杂志角",对于建筑周围营造出热闹的氛围。

注重站立阅读的同时也没有忽视休闲阅读。沙发、长椅等小品随处可见。而且专为书店设计了适合放置多本书籍的大扶手。另外,三层的画本区使用了彩色地毯,孩子们可以一边玩耍一边读书。设计师以三层的中心柜台为起点向外放射设计了书架,那么站在中心柜台可以望见整个楼层,不留死角,因此可以防止书籍被盗窃,或者留意孩子的安全。在柜台周围放置了沙发,坐在沙发上的客人同样可以帮助店员留意孩子和书籍。豆子主题随处可见,一、二两层柜台的侧面,园林的地面砖,独立广告板以及标识都有豆子的影子。

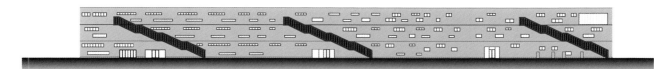

Interior and exterior expansion drawing
室内外展开图

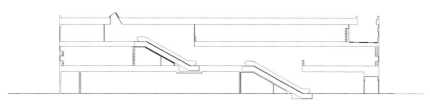

Section 剖面图

1. Ground floor cashier
2. 2nd floor cashier
3. The children used area
4. Showcase and stand reading area

1. 一层收银台
2. 三层收银台
3. 儿童区
4. 书架和站立阅览区

Ground floor plan 一层平面图

1. Entrance
2. Showcase
3. Café
4. Reading cornor
5. Escalator
6. Bathroom

1. 入口
2. 展示区
3. 咖啡厅
4. 阅读角
5. 电梯
6. 洗手间

1. Escalator
2. Stand reading area
3,4. Children reading area

1. 自动扶梯
2. 站立阅览区
3、4. 儿童阅览区

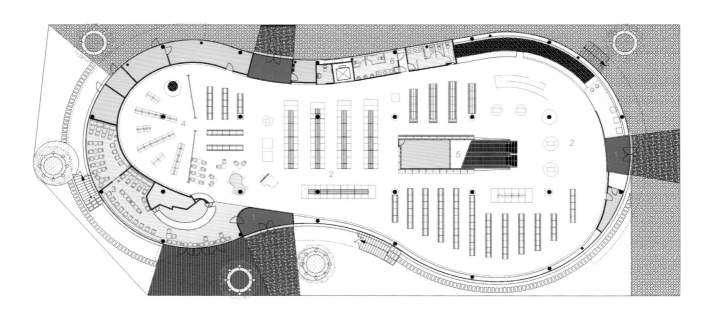

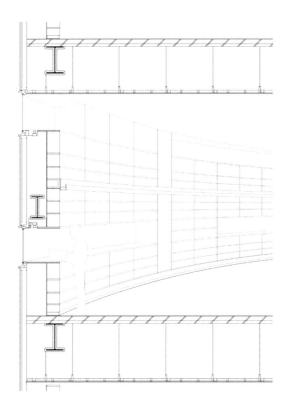
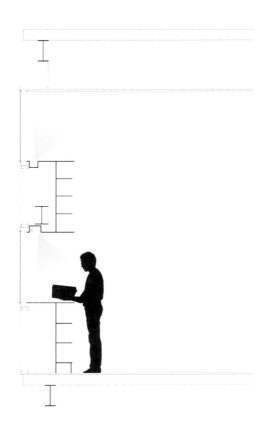

Stand reading drawing
站立阅览区图示

Completion date: 2007
Location: Bolzano, Italy
Designer: monovolume architecture + design (Patrik Pedó_Juri Pobitzer)
Photographer: Oskar Da Riz
Construction area: 1,250sqm

Blaas General Partnership

BLAAS 博尔扎诺

Key words: A Full Exhibition by Glass House with Office Area Shaded in Louvres
Three facades of the building are enclosed with glazing, which help to achieve high level of transparency and provide natural light. The "glass house" design provides a maximum of visibility and transparency to the electro-mechanics that the store sells. Since the client wants to use the first floor as office area, the designer applies louvre screening system outside the glazing to protect, maintaining unity of the whole building.

主题词："玻璃房"的全面展示搭配"百叶"掩映下的办公区
该项目的外立面有三面自下而上完全由玻璃围合，实现了大尺度的透明性，并提供了极佳的采光条件。由于该商店出售的是机械设备，这种"玻璃房"式的设计能够最大限度的陈列、展示所售商品。业主想使建筑的二层空间用于办公，所以设计师在玻璃立面外采用了"百叶窗"式的防晒系统进行保护，保持了整个建筑的统一性。

Design Points:

The company Blaas in Bolzano is specialised in electro-mechanics. In the new head office the company presents its new product range and offers repair service.

On the ground floor of the building there is the sales division, on the first floor the exposition area and the repair shop. All administration offices are located on the second floor. The overall impression of the structure is a homogenous and closed building. Nevertheless, there exists a separation between the public and the private sector. The client can perceive this clear and formal internal division already from the outside.

The glass façade on the northern side provides a maximum of visibility and transparency to the exhibition and sales area. The private spaces such as repair offices, stockrooms and offices have their façades exposed to the south, east and west which are protected with a sun screening system.

In order to establish an optimal relation between nat-

ural light, development and planning of spaces there has been created a luminous entrance hall in the centre of the building with an inner courtyard. This green open spot permits the administrative sector of the second floor to receive ample natural light and at the same time it generates a protected, quiet recreation area for the staff.

The modern Blaas headquarters are the result of a consistent fusion between customer's needs, context and utilisation. As a supplier for solid construction companies, Blaas needed huge yet flexible spaces to give maximum prominence to large items. Consequently, a fairfaced concrete stucture

1. South view
2. Internal roof garden
3. Main entrance with staircase

1. 建筑南侧外观
2. 屋顶花园
3. 主入口和楼梯

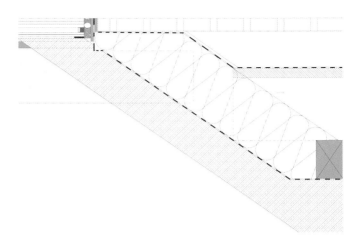

来自博尔扎诺的BLAAS是一家专业从事电子机械的公司。在新的总部办公大楼，品牌展示了新产品，并提供维修服务。

大楼的底层是销售区。一楼是展会区和维修店。所有的行政办公室都位于二楼。结构给人的整体印象是一个具有同质性的、封闭的建筑。然而，公共部门和私人部门被分离开来。客户从建筑外部就可以明显地感觉到内部分区。北侧的玻璃幕墙上最大程度地为展示和销售区域提供了可视度和透明度。维修室、库房和办公室等私人空间的外墙朝向南、东、西三个方向，他们被太阳光遮蔽系统保护着。

为了在自然光和空间的发展和规划之间建立一个最完美的关系，我们已经在大楼的中心建了一个明亮的入口，并带有一个内庭院。这处环保设计使二楼的行政部获得了充足的自然光线，同时为员工提供了一个受保护的、安静的休闲区域。

1-3. Detail image
1~3. 细部设计

Detail drawing
细部图

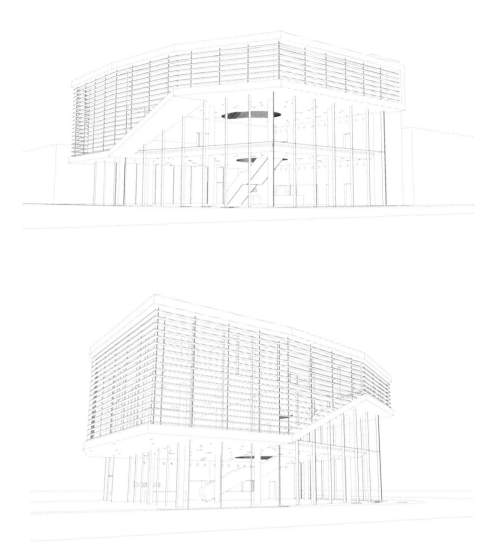

3D elevations
三维立面图

with the least possible distracting pillars was sought in collaboration with the structural designer, whilst a full glass façade extending across two floors, developed in teamwork with glass engineers, acts as a wrapping for the resulting open showcase. A five-metres overhang serves as a canopy for the clients entrance on the ground floor and as an employee outdoor area for special workshop activities on the first floor. The entrance area features an inviting free standing stairway with an organic stairway opening and a skylight above. This skylight allows excellent two-sided illumination of the office work places. The massive office decor construction style present enough thermal mass to allow the construction to absorb heat during the day and to release it during the night.

现代BLAAS总部是在消费者的需求、环境和用途之间不断融合之下产生的。作为建设公司的供应商，BLAAS需要巨大而灵活的空间，以便给予大型项目最大的重视。因此，（砌砖）清水面的混凝土结构包含与结构设计师的协作中探索出的最不易分散的支柱，同时全玻璃幕墙延展跨越两层楼。幕墙是与玻璃工程师合作开发的，充当开放式陈列橱窗的外"包装"。

底层有一个5米高的悬垂部分，被用作入口，同时一楼有一个雇员室外活动区域，可以在进行专题研讨会时使用。入口区有一个舒适的楼梯，顶端有一个天窗。天窗使办公区拥有双面照明。大量的办公装饰建筑风格呈现足够的热质量，允许建筑在白天吸收热量，并在夜间释放。

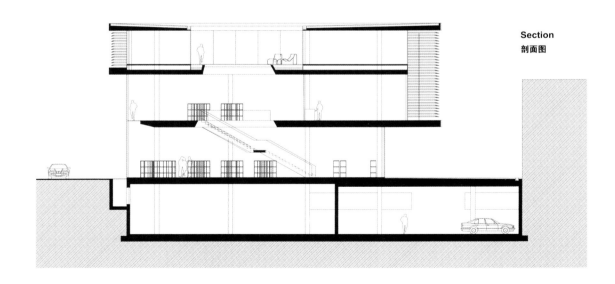

Section
剖面图

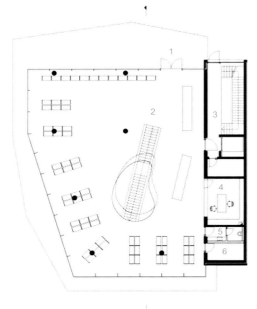

Ground floor plan

1. Entrance
2. Sales/shop
3. Staircase
4. Office
5. Bath/WC
6. Store room

一层平面图

1. 入口
2. 售货区
3. 楼梯
4. 办公区
5. 洗手间
6. 仓库

1. First floor sales area
2,3. Detail of the main staircase
4. Main staircase

1. 二层销售区
2、3. 主楼梯的细部设计
4. 主楼梯

CHAPTER 2 **Cases Study** | Detached Commercial Architecture

1. Ground floor sales area with main staircase
2,3. Offices on the first floor
4. Staircase
5. Toilet

1. 一层销售区和主楼梯
2、3. 二层办公室
4. 楼梯
5. 洗手间

CHAPTER 2 **Cases Study** | Detached Commercial Architecture

Completion date: 2011
Location: Tokyo, Japan
Designer: Astrid Klein, Mark Dytham/ KleinDytham architecture
Photographer: Tomoko Ikegai
Construction area: 2322,26sqm

Daikanyama T-Site 代官山茑屋书店

Key words: Using Façade Design to Highlight the Brand
The bookstore's grid form is organised with several T structures joint together. The T structures stands for Tsutaya's logo, which is easy to understand. The white bricks with T shape on the exterior walls are interesting and practically promote the brand. The building is divided into three levels. Grids are connected with the main structure. The designer also introduces the grid patterns into the interior design skillfully, emphasising the logo's image.

主题词：彰显品牌的立面设计
这间书店的外表面是格子状的造型，由多个相互联合的"T"字型结构排列组合，类似茑屋书店的LOGO，极具代表性，该LOGO易读且形式简洁。"T"字型的外墙白砖有趣而实用地起到了宣传的作用。该项目共有三层，格子状结构与建筑整体的组成部分相连，设计师还巧妙的将建筑外立面的格子结构延续到室内立面的装饰元素，使这种纹理深入人心。

Design Points:

KDa's new Daikanyama T-Site is a campus-like complex for Tsutaya, a giant in Japan's book, music, and movie retail market. Located in Daikanyama, an up-market but relaxed, low-rise Tokyo shopping district, it stands alongside a series of buildings designed by Pritzker Prize-winning architect Fumihiko Maki.

Drawing on all KDa's design skills – architecture, interior, furniture and product display – the project's ambition is to define a new vision for the future of retailing. This selection of goods is intended not to be exhaustive but stimulating – sections include art, architecture, cooking, cars, design, history, and literature. Each section is run by a concierge – like their intended customers they are all over 50, and both have expert knowledge of their subject area and can provide other services. The project also merges the worlds of new and old media. Movies, for example, can be bought, rented, or downloaded, and while iPads are on hand throughout the store as guides to the stock on offer,

The pavilions contain retail space on the lower floors with accommodation above. Internally, KDa didn't

want a slick "department store" feel, and so selected materials such as aged timber flooring to create a relaxed space – part university library, part warehouse. The three pavilions are linked by an organizational spine – called the "magazine street" – which passes through interior and exterior, linking the three buildings together. To reinforce its presence, the magazine street's shelving, flooring and slatted ceiling are all timber, even where it runs outside under the bridge. Elsewhere, a stone floor creates continuity between interior and exterior while an open ceiling gives a warehouse feel (big lighting lanterns hang down to subtly mask the buildings' innards) and signage designed by graphic maestro Kenya Hara was printed on perforated metal to create more openness and visibility.

1. Exterior view and surrounding
2. The middle pavilion
3. Shopfront

1. 建筑外观及周边环境
2. 中央馆
3. 店面

1. Magazine street
2. Book showcase
3. Book category labels

1. 杂志街
2. 图书展示
3. 图书分类标签

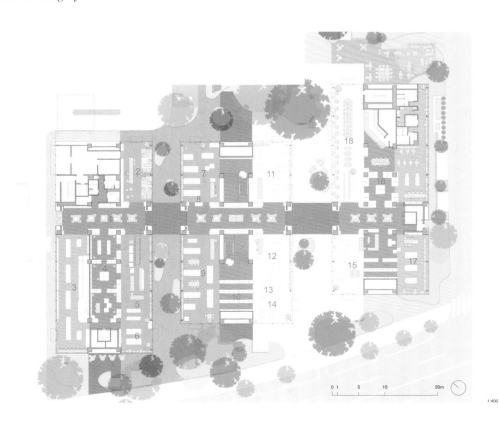

Site plan

1. Magazine street
2. Information
3. Convenience store
4. Philosophy
5. Literature
6. Picture story book
7. Architecture
8. Interior
9. Car & Motorbike
10. Camera & Watch
11. Design
12. Art
13. Photograph
14. Fashion
15. Cooking
16. Travel
17. Stationary
18. Café

总平面图

1. 杂志街
2. 信息中心
3. 便利店
4. 哲学类图书
5. 文学类图书
6. 图画故事书
7. 建筑类图书
8. 室内类图书
9. 汽车摩托类图书
10. 相机钟表类图书
11. 设计类图书
12. 艺术类图书
13. 摄影类图书
14. 时尚类图书
15. 烹饪类图书
16. 旅行类图书
17. 固定设施
18. 咖啡厅

Within this organisational and material framework, each of the boutique spaces has its own character – the shelving in the literature section is tightly packed to evoke Tokyo's atmospheric Jimbocho second hand book district, while in other spaces overhead shelves are used to make them feel more intimate. Other facilities include a café, an upscale convenience store, and the Anjin Lounge. Located in the center of the complex, this lounge includes a bar, a performance space, a collection of artworks and rare books for sale that visitors can enjoy as they eat, drink, read, chat, or relax. Visitors to the lounge can also browse an amazing world magazine archive that includes beautifully bound collections .

Despite its design innovations, this was not a big budget project – it was low cost and produced extremely quickly. The whole project was completed in 20 months, with construction taking just 11 months despite disruptions to Japan's construction industry caused by the Tohoku earthquake and tsunami. Their scheme triumphed by tackling both explicit branding issues and architectural subtleties in every aspect of the building. Merging the digital and analogue worlds, and providing for both sophisticated tastes and simple curiosity, KDa's design is intended to allow a new retail paradigm to emerge.

KDa 事务所设计的新代官山茑屋书店是一个校园式综合体。它是日本一个大型的书籍、音乐、电影零售市场。书店位于代官山，这里虽然消费高档，但是氛围轻松，属于东京低层购物区，周围的建筑由普利兹克建筑奖获奖建筑师槙文彦设计。

项目充分展示了 KDa 事务所的设计水平——无论是建筑、室内、家具，还是展区都表明该项目的目标是展现一个新的未来零售空间。选择商品的目的不是详尽的，但是充满刺激，产品内容涉及艺术、建筑、烹饪、汽车、设计、历史和文学。每个部分都有专人服务，

1. Book showcase
2. Upper floor and dormer
3. Reading area

1. 图书展示
2. 二楼空间和天窗
3. 阅读区

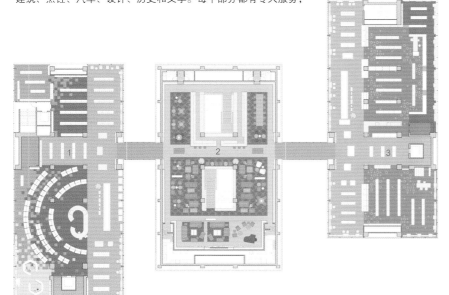

Ground floor plan

1. Movie
2. Library & Lounge
3. Music

一层平面图

1. 电影
2. 图书室&休息室
3. 音乐

就像喜欢他们的客户一样,他们也都超过50岁,都有其学科领域的专业知识,并能提供其他服务。该项目还是个融合了新老多媒体的世界。例如如果你想看电影,你可以买,可以租,也可以下载。还可以将ipad拿到手里随时关注手头上的股票报价。

零售区在较低的楼层,上面是住宿空间。KDa并没有想把内部营造成一个漂亮的"百货商场"的感觉,所以选择的材料都是有年头的木材,以创造一个轻松的空间,即一半像大学的图书馆,一半像仓库。三个展馆由一个脊柱组织相连,被称为"杂志街"——贯穿内部和外部,将三座建筑连接在一起。要加强它的存在,杂志街的货架、地板和天花板板条均是木材所制,即使是在外部桥下的部分也是如此。一个石头在内部和外部之间,这里的开放天花板给人一种仓库的感觉(亮着的灯笼似天窗高垂着,巧妙地掩盖了建筑物的内部结构),平面设计大师原研哉设计的标志被印在穿孔的金属上,增加开放性和可见性。

在这样的组织和物质框架内,每个精品空间都有自己的特色:文学区的架子挨得很紧密,以唤起人们对东京的神保町二手书的关注,而其他地方都使用的高架,使读者感到更贴心。其他设施包括一间咖啡厅、高档便利店和Anjin休息室。休息室坐落在中心位置,设有一间酒吧,一个表演空间,艺术品和珍贵书籍收藏区,顾客可以边欣赏边吃喝、阅读、聊天或放松。在休息室,游客还可以浏览到奇妙的、装订精美的世界杂志档案。

虽然在设计上有所创新,但这项目的预算并不大,不但成本低,完成速度还非常快。尽管日本的建筑行业受日本东北大地震和海啸的影响很大,整个项目只历时20个月,施工期只有11个月。计划成功实施的同时明确了品牌在各个方面的问题和建筑的每个细微之处。项目通过合并数字和模拟世界,同时满足了复杂的口味和简单的好奇心。KDa的设计目的正是为了打造一个新的零售设计范例。

1. Reading area
2. The bar
3. Magazine showcase
4. Reading area
5. Stair

1. 阅览区
2. 书吧
3. 杂志展示
4. 阅览区
5. 楼梯

Completion date: 2011
Location: Nagaoka, Japan
Designer: M Sc.Ton Smulders
Architect: Oki Sato, nendo
Photographer: Masaya Yoshimura
Construction area: 700sqm

HALSUIT 春山西装店

Key Words: A Sharp Building Shape Consist of two Geometric Blocks
As a men's suit boutique, the building is formed with a reversed L and a rectangular block. The building looks tough and sharp and the two modules interlock together. The light belts on the edges highlight the perspective two blocks, reminiscent of masculine beauty. The Logos on the front and side faces of the shop look direct and brief. The building style is perfected combined with the theme of a men's suit shop.

主题词：几何模块组成棱角分明的建筑造型
本项目是一家男装精品店，建筑造型由一个反"L"形几何模块和一个长方体组合而成，建筑造型硬朗、交错变化、棱角分明，在两个建筑模块的边缘采用了灯带照明，更凸显出两个体块的立体感，使人马上就能联想到男人的阳刚之美，建筑正面和侧面的店铺LOGO直接明了、简洁大方。这种建筑风格与男装店的主题完美的结合在了一起。

Design Points:

This is a concept shop for high-volume men's suit retailer Haruyama. Haruyama offers its customers a dizzying number of options for suits, shirts and ties.

Variety of choice is one of Haruyama's strengths, but customers can also be overwhelmed by the number of options, and find it difficult to choose the best matches for their suit. Families can get bored, and the overall atmosphere isn't always as conducive to relaxed, pleasant shopping as Haruyama would like. The designers' first decision was to move the fitting rooms from the edge of the shop floor to its centre. The designers used the front exterior walls of the fitting rooms as showcases for different ways of coordinating the suits, and installed a counter with magazines and television for friends and families. Matchable accessories are arrayed around the area, making it into a focal point where shoppers can develop an image of the suit they'd like to have.

Most suits are worn to the office, so the designers used lighting reminiscent of desk lamps, and shelving in the style of office storage units to create the right scene. The designers replaced posters with LCD screens, and transformed the sales area for shirts into a server room. Shoppers make their purchases

at a "reception desk", and lounge and conference room-type spaces help shoppers to imagine their own work styles as they select their suit. The space reflects Haruyama's brand concept, that men should define and enjoy their personal working style. The designers used louvers that change colour depending on the angle from which they're viewed for the shop exterior, so that the image of the shop varies, depending on the direction from which drivers approach.

1. The shopfront
2,3. The exterior view in night

1. 店面
2、3. 建筑外观夜景

1. The shirt showcase 　1. 衬衫展示区
2. The suits showcase 　2. 套装展示区

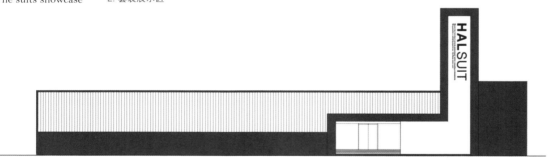

1:400

Elevation 立面图

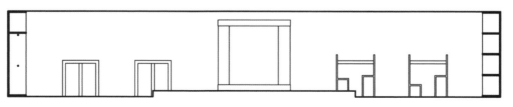

Section 剖面图

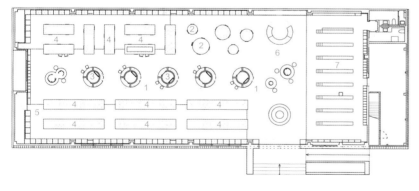

1:400

Ground floor plan
1. Display Space
2. Display Table
3. Fitting Room
4. Hanger
5. Mirror
6. Register Counter
7. Shelf

一层平面图
1. 展示区
2. 展示台
3. 更衣室
4. 衣架
5. 镜子
6. 收银台
7. 货架

这是日本高端男装品牌春山的一家概念店。春山西装专为消费者提供各种各样的套装、衬衫和领带。

品种繁多是春山西装的优势之一，但是消费者可能会感到眼花缭乱，难以选择最好的套装。这样一来，消费者的陪同人员会感到无聊，很难实现春山西装所希望看到的轻松、愉快的购物体验。因此，设计师首先考虑将试衣间从店铺边缘移到了店铺中心。他们利用试衣间的外墙来展示各种西装搭配，并且专门设计了摆放杂志和电视的柜台来取悦陪同人员。试衣间四周摆放着配饰，将整个区域打造成了挑选西装及配套产品的焦点。

大多数西装都是在办公室穿着的，所以设计师选用与台灯相似的灯光以及与办公室储物柜相似的衣架来营造类似的场景。设计师以液晶屏幕替代了海报，并且将衬衫售货区改造成了服务机房。消费者在"前台"挑选商品，休息室和会议室风格的空间能帮助他们想象自己穿上西装工作的样子。店铺空间反映了春山西装的品牌概念——男士应当明确并享受自己的个人工作风格。设计师在店铺外采用了能够变换色彩的百叶窗，使店铺的形象随着观看角度的不同实现了变化。

1. The cashier
2. The bathroom
3. The suits showcase
4. The desk style showcase

1. 收银台
2. 洗手间
3. 套装展示区
4. 桌面展示

CHAPTER 2 **Cases Study** | Detached Commercial Architecture

Completion date: **2007**	Designer: **Andrej Kalamar, Studio Kalamar**	Site area: **6,900sqm**
Location: **Ormož, Slovenia**	Photographer: **Studio Kalamar**	Construction area: **11,120sqm**

Retail Centre Holermuos
霍尔莫斯购物中心

Key words: Alternative Treatments of Various Materials and Details
The project features minimal materials: concrete, printed glass and black metal panels are alternatively used. Each material is treated with details: the front concrete elevation is pierced with various triangles with round edge, looking lively; the glass clad elevation is highlighted with silver coloured print on glass and thin vivid colour verticals; the other two elevations are simply clad in black metal panels.

主题词：多种材料与细节的交替处理
该项目的特征是采用极简材料——混凝土、印花玻璃和黑色金属板交互使用，并在每种材料上都进行了细节的处理，混凝土的正立面上穿透了数个大小各异的圆边三角形，颇为生动；与混凝土相接的玻璃里面上带有箭头形状的银色印花，并有色彩跳跃的立柱装饰；黑色金属板则较为简洁的包裹了建筑的另两个立面。

Design Points:

The centre is located at the edge of town, next to a public park. The three-storey west wing houses retail spaces and offices, the principal one-storey volume houses a supermarket. Back and side elevations that face green areas are characterised by their minimalist colour and texture scheme, combining dark gray metal panels and visible concrete. Front elevations, facing the city, introduce several visual accents: construction of the glass clad eastern elevation is highlighted by a series of thin vivid colour verticals, while pixelated silver coloured print on glass disperses morning sun. The horizontal composition of lower volume's glazed northern elevation is accentuated by the two bright coloured entrances, connected by the green overhang.

Interior is characterised by a neutral coloured ceramic floor, interchanging matte and glossy tiles. The entrance atrium is opened to the upper floor. Its principal feature is an elevated café with a pronounced overhang. A series of conical cupolas on the ceiling provides ample natural light.

Changes during design process:
Minor changes happen during most creative processes, influenced by various factors that come up during the work process, but the basic idea mostly stays as it was conceived in the beginning, based on input of information from different fields that will have input on the design of the building.

1. North elevation in combination of glass and visible concrete
2. North and east elevations: glass, printed glass, visible concrete
3. Overhang in green accentuates and protects the entrances

1. 建筑北立面结合了玻璃与混凝土结构
2. 北立面和东立面：玻璃、印花玻璃、混凝土相结合
3. 绿色的雨篷突出了入口，并起到了防护作用

CHAPTER 2 **Cases Study** | Detached Commercial Architecture

1. View front the south: building faces the lake and the park
2. Eastern elevation in black metal panels
3. Interior corridor: sunlight reflects the printed pattern from the glass onto the floor
4. Southeast corner: façade in combination of concrete and metal panels

1. 建筑正面和南侧：建筑朝向湖泊和公园
2. 东立面采用了黑色金属板
3. 室内走廊：阳光透过玻璃投射到地面上，留下印花图案的阴影
4. 建筑东南角：外立面结合了混凝土和金属板

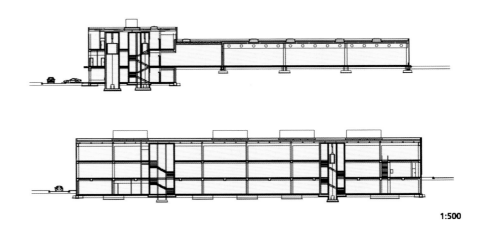

1:500

Sections 剖面图

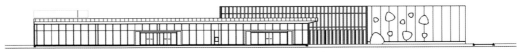

North elevation 北立面图

购物中心位于城镇边缘,靠近一座公园。三层高的西楼内是零售店和办公区,单层主体结构内则是超市。建筑朝向绿地的后面和侧面以简洁的色彩和外墙结构为特色,结合了浅灰色金属板和露石混凝土。建筑正面朝向城市,引入了若干视觉元素:东面的玻璃外墙上添加了鲜艳的彩条,玻璃上的银色像素化印花分散了晨光。低层结构北面的玻璃外墙上点缀着两个亮色入口,二者由绿色的雨篷连接。

室内地面采用了中性色彩,以亚光和亮光的陶瓷地砖交替平铺。入口中庭直通顶楼,最显著的特色在于一家架高的咖啡厅,咖啡厅的悬臂结构引人注目。天花板上的一系列圆锥形天窗为室内提供了自然光。

设计过程中的变化:
在创意过程中,由于各种因素的变化,设计做出了一些小改动,但是项目的基本理念一直保持不变:来自各个方面的信息都体现在了建筑的设计上。

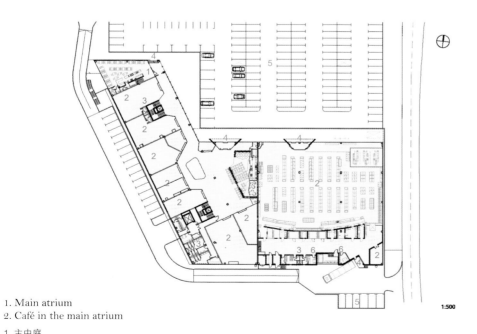

Floor plan

1. Restaurant
2. Shops
3. Communications
4. Entrances
5. Parking
6. Secondary spaces

平面图

1. 餐厅
2. 店铺
3. 流通空间
4. 入口
5. 停车场
6. 辅助空间

1. Main atrium
2. Café in the main atrium

1. 主中庭
2. 主中庭中的咖啡厅

Completion date: 2009
Location: Bangkok, Thailand
Designer: David Mayer, Boriphon Suwattana/Whitespace
Photographer: Wison Tangthunya
Construction area: 3,000sqm

J&B Beauty World   J&B 美丽世界

Key words: A Harmonious Combination of Pierced Screen and Ship Structure
The project looks well-proportioned and undulating both horizontally or vertically, full of dynamics and energy. A sculptural sun screen perforated by high tech machinery drapes over the façade. A ship's bow formed with glass makes the building seems to a real ship. The pierced screen and glass ship bow combines solidness and transparency while shows the products moderately and mysteriously, achieving a balance effect.

主题词：镂空屏障与类船结构的和谐组合
该项目无论从横向还是纵向来看都是错落有致、高低起伏的，这使得整个建筑极具动感和活力。设计师运用高科技机械镂空而成的如雕塑般的屏障垂直挂在建筑之上，又建造由玻璃组成的似轮船船头的造型宛如巨轮呼之欲出。镂空屏障与玻璃船头结构实现了虚实的搭配，既保有了一定神秘感又适度的展示了所销售的产品，一虚一实，张弛有道。

Design Points:

Over the last 29 years, the J&B company had quietly become the most important retailer of products to the salon industry in Thailand. Operating from modest headquarters in Bangkok's Pratunam area in the past years, the need to relocate due to development in the area prompted owners K.Thatinee Sawasdee and K. Saijai Sarikanont to seize to opportunity to create a proper headquarters that accurately reflects their leadership position in the industry. The new building is located at Petchaburi road near Chidlom in central Bangkok.

J&B decided to work with Whitespace.
Whitespace design solution carefully balances branding, architecture, and retail interiors. They describe the architecture as purely modernist; simplicity expressed with excitement and creativity. The building exterior includes a 6 story high sculptural sun screen "draped" over the southern facade of the building. This screen was perforated by high tech machinery with a custom designed pattern by Whitespace graphic designers. Another example is the multi-storeyed glass showcase window that dramatically projects from the main body of the building like the bow of a ship.

J&B's new facility includes 1,000m² of retail on 4 floors including every conceivable product needed by professional salons. Whitespace devised a display and environmental graphic system bringing colour and drama to each floor while making it easy for customers to shop. Floor 1 is a convenience area presenting the range of most popular and products. Floor 2 features hair products and cosmetics, while floor 3 highlights spa products and salon furnishings. Furnishings continue to the 4th floor which also contains a seminar room with all the necessary facilities to present hair product and styling demonstrations. Upper floors contain administrative and private facilities.

1. Exterior view of the multi-storied glass showcase window
2. Exterior view from the main body of the building like the bow of a ship
3. Showcase window

1. 多层展示橱窗的外观
2. 建筑主体外观看起来像一艘轮船
3. 橱窗

CHAPTER 2 **Cases Study** | Detached Commercial Architecture

在过去 29 年中，J&B 公司已经在不知不觉中成为泰国最重要的沙龙行业产品零售商。在过去几年中，J&B 总部低调地设在曼谷水门区。搬迁是由于品牌在该地区的发展提示 K.Thatinee Sawasdee 和 K. Saijai Sarikanont 抓住机会，建立一个更合适的总部以准确地反映其在行业中的领导地位。新大楼位于曼谷市中心 Chidlom 附近的 Petchaburi 大路。

J&B 公司决定同 Whitespace 合作。Whitespace 的设计方案注重平衡品牌、建筑和室内装潢。他们将建筑描述为纯粹的现代主义，用兴奋感和创造性表达一种简洁。建筑包括一个 6 层楼高的雕塑般的太阳屏，"披"在大楼的南立面上。设计师利用高科技穿孔在此屏幕上做出客户要求的图案。另一个例子是多层玻璃橱窗，使得建筑的主体部分犹如船首。

J&B 的新楼一共四层，零售区面积达到 1000 平方米，包括专业沙龙需要的所有产品。Whitespace 设计了一个展示和环境图案系统，每一层都使用不同的颜色，以方便顾客购物。一楼作为最方便的区域，用来呈现最流行的产品。二楼主营护发产品和化妆品，而三楼的亮点是水疗产品和沙龙家具。四楼的高级体验室有很多专业设施，用来呈现护发产品的效果和一些造型示范。最上层是行政和私人设施。

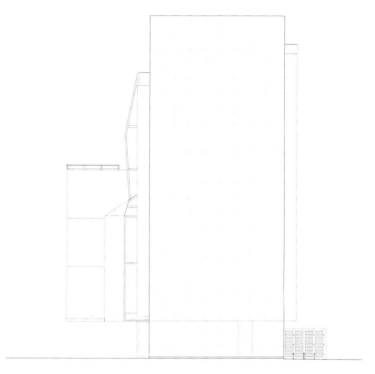

Elevation 立面图

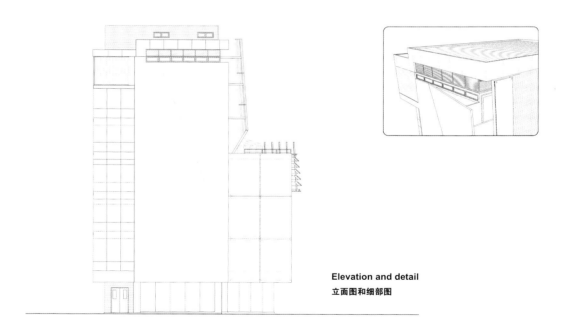

Elevation and detail
立面图和细部图

1. Exterior view interior light glowing at night
1. 建筑内部的灯光在夜晚闪耀

1. Interior view of the perforated screen
2. Stair
3,4. Product feature area on 1st floor with triple height ceiling space

1. 镂空网内部
2. 楼梯
3、4. 二层产品展示和三层楼高的天花板

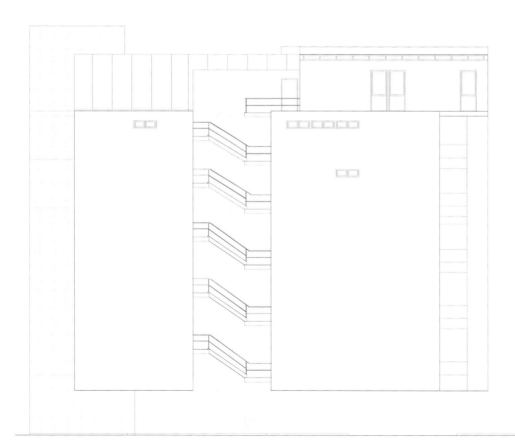

Elevation 立面图

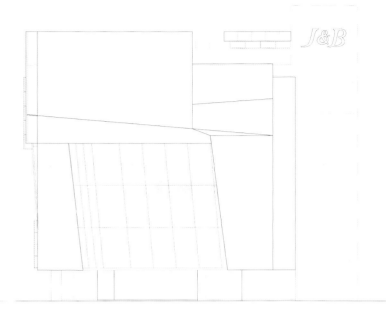

Elevation 立面图

CHAPTER 2 **Cases Study** | Detached Commercial Architecture

1. The ground floor is a convenience area
2. Cashier
3. The ground floor is a convenience area
4. Stair

1. 一层空间十分便捷
2. 收银台
3. 一层空间十分便捷
4. 楼梯

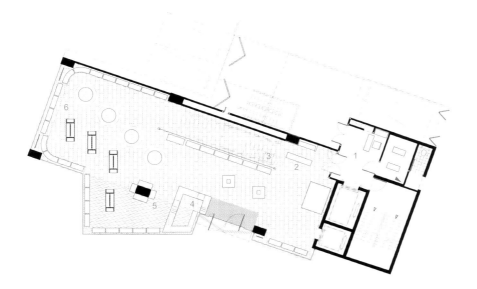

Ground floor plan
1. Security
2. Service counter
3. Storage
4. Cash desk
5. Cosmetic
6. Hair colour

一层平面图
1. 保安室
2. 服务台
3. 仓库
4. 收银台
5. 化妆品
6. 染发区

Completion date: **2008**
Location: **Beverly Hills, USA**
Designer: **Standard, Jeffrey Allsbrook**
Photographer: **Benny Chan**
Construction area: **464.52sqm**

James Perse Flagship Store
詹姆斯珀思旗舰店

Key Words: A Delicate and Concise Three-dimensional Shopfront
The project uses concise design and skillfully takes advantage of materials and light, creating a unique new store in the block. White basic colour is matched with wood sliding shutter-doors. Open the sliding doors, a delicate show window will be presented. Structurally, the set back entry and recessed upper storey loggia create a three-dimensional and mysterious effect, catching passers-by's eyes.

主题词：精致简约的立体立面
该项目运用了简约的设计手法，巧妙的利用了材料和光线的作用，设计出在这个街区具有特点的新店。白色的主体建筑底色，搭配棕色的木质百叶滑门。当滑门拉开，设计精致的橱窗便呈现出来；在结构上，将店铺入口和二层窗户设计凹陷于建筑，建筑更显立体、神秘，吸引着路人光临。

Design Points:

The Beverly Hills flagship store is the eighth retail store in over a dozen projects that Standard has designed for James Perse since 1998. Other significant projects include the original James Perse store on Melrose (2003), stores in Malibu (2005), Las Vegas (2006), San Diego (2006) and two in New York City (2005 and 2006). Standard also designed Perse's Los Angeles residence, which was completed in 1999.

From the original Melrose store to this new flagship, Standard has employed their refined and minimal style to create distinctive boutiques for James Perse's expanding lines of casual luxury clothing and accessories. Throughout their work, Standard combines a disciplined architectural approach with sensitivity to light and material. Their collaboration with James Perse has yielded a body of work that stands out in the world of retail design.

With its careful attention to detail, proportion and its elemental use of materials, the Canon Drive store maintains continuity with previous James Perse stores even as it expands to accommodate the growth of Perse's luxury label. Organised around a garden

courtyard that features a custom designed steel fountain, the boutique captures the comfort of a Southern California domestic setting through generous use of natural light and blurring of indoor and outdoor spaces, much like Standard's residential projects.

Standard radically altered the existing two-storey building with a new storefront that confidently announces the arrival of James Perse. The bold but minimal façade transforms the north end of sleepy Canon Drive. Distinct from adjacent retail stores and restaurants, the façade features floor to ceiling sliding teak shutter-doors and a recessed upper storey loggia. Entry to the store is set back from the sidewalk, veiled behind the shutters. These shutters slide horizontally to reveal or con-

1. Front façade with closed shutters
2. Entry ramp
3. Front façade with closed shutters

1. 外立面和封闭的百叶窗
2. 入口坡道
3. 外立面和封闭的百叶窗

CHAPTER 2 **Cases Study** | Detached Commercial Architecture

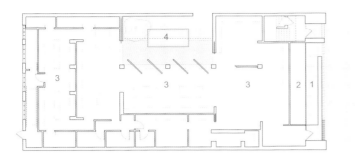

Ground floor plan   一层平面图
1. Ramp   1. 坡道
2. Window   2. 橱窗
3. Retail   3. 零售区
4. Courtyard   4. 庭院

詹姆斯珀思的比弗利山旗舰店是自1998年以来，标准设计公司为詹姆斯珀思公司所设计的第八家零售店。其他富有影响力的项目包括梅尔罗斯的首家詹姆斯珀思店（2003年）、马布里店（2005年）、拉斯维加斯店（2006年）、圣地亚哥店（2006年）以及两家纽约店（2005年和2006年）。标准设计公司还为珀思先生设计了他在洛杉矶的宅邸，整个项目于1999年竣工。

从梅尔罗斯的第一家零售店到这家新旗舰店，标准设计公司以精致简约的风格为詹姆斯珀思品牌打造了独特的精品店，对其销售轻奢品牌服饰和配饰起到了促进作用。通过努力，设计公司将优质的建筑与灯光和材料巧妙地结合在一起。他们与詹姆斯珀思的合作远远超过了零售设计的范畴。

这家旗舰店对细节、比例和材料运用都格外重视，既保持了与前几家詹姆斯珀思零售店的一致性，又彰显了品牌的奢华气度。店铺中央是一个花园式庭院，以定制的钢铁喷泉为特色。店铺通过运用大量自然采光和模糊室内外的界限，展现了南加州的家居风格的舒适感，与标准设计公司的住宅项目十分相似。

设计师对原来的双层建筑进行了全面改造，形成一个全新的店面。大胆而简约的外立面改变了佳能大道北端沉闷的氛围。与周围的零售店和餐厅不同，项目的外立面采用了柚木百叶落地门，还在二楼使用了嵌入式凉廊。店铺入口缩在人行道后方，隐藏在百叶门之后。这些百叶门可以水平拉动，有选择地隐藏或展露店铺橱窗和入口。分层的外立面和嵌入的入口形成了一个过渡区，起到了与古典宅邸的门廊相似的效果。这让入口空间模糊了室内与室外、公共与私人的概念，把它们融合在一起。店铺在注重内部设计的同时，也对周边城市环境起到了点缀装点的作用。

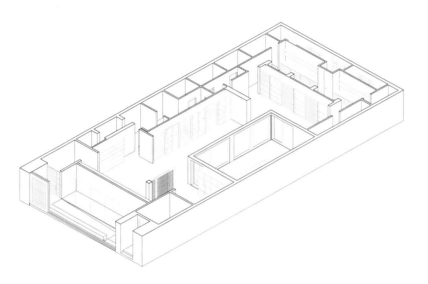

Axonometric view
轴测图

ceal the long shop window and entrance. The layered façade and set back entry yields a transitional zone that functions much like a portico characteristic of a classical palazzo. This protected entry space blurs the boundary between inside and outside, public and private, seamlessly merging the two without compromising either. While the store is predominantly focused inward, it gives something back to its urban context by way of this semi-public space that is a product of a layered façade.

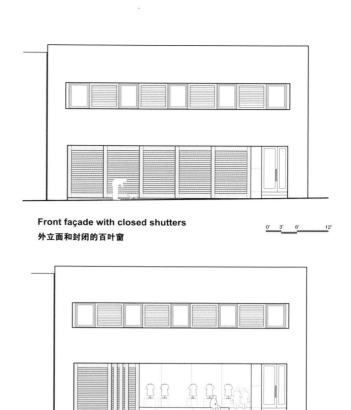

Front façade with closed shutters
外立面和封闭的百叶窗

Longitudinal section
纵剖面

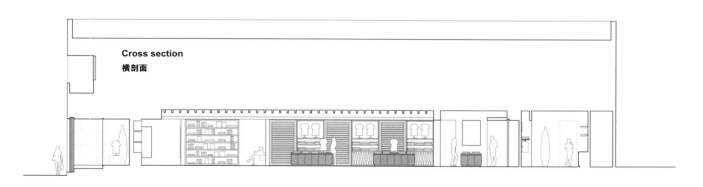

Cross section
横剖面

1. Store interior adjacent to courtyard
2, 3. Store interior

1. 靠近庭院的室内设计
2、3. 店铺内部

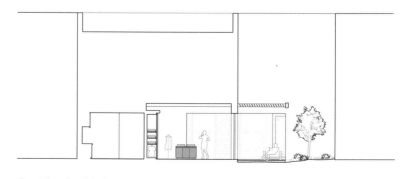

Front façade with closed shutters
外立面和封闭的百叶窗

CHAPTER 2 **Cases Study** | Detached Commercial Architecture

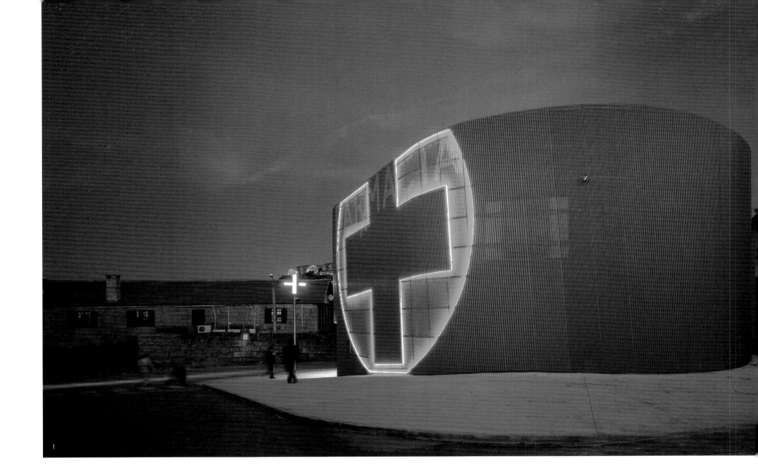

Completion date: 2012
Location: Vila Real, Portugal
Architect: José Carlos Cruz - Arquitecto
Photograher: FG+SG – Fotografia de arquitectura
Construction area: 522sqm

Lordelo Pharmacy 洛尔德罗药店

Key words: A Clean and Concise Abstract Form
Located in a block without unified image or style, the two-level building features an abstract and modern form. It is a cylinder without any windows or openings except for its entrances, completely breaking the typical image of a pharmacy. The façade is covered with dark folded aluminium panel, forming a unique texture. A whole elevation on one side of the building is covered with a large cross with green lighting belt on its edge, which creates a clear identity for the pharmacy, especially at night.

主题词：简洁明了的抽象造型
该项目所处的街区没有统一的形象或者风格特点，这个二层建筑整体造型抽象、现代——呈椭圆形，除了建筑的出入口，表面封闭、没有任何窗户或开口，一改往日人们对药店的印象。外立面被深色皱褶铝板所覆盖，颇具质感。建筑一边的整个立面都被一个巨大的十字符号所覆盖，十字符号边缘用绿色灯带装点，药店的"身份"尽显，在夜半时分格外耀眼。

Design Points:

The pharmacy is located in Vila Real, in the centre north of Portugal and is part of a peripheral zone of the city where the environment does not have a consolidated and uniform image. In the absence of external references, it was chosen to create a building with an abstract and neutral character, reinforced by the absence of openings. With oval shape footprint, the two floors are fully aluminium coated corrugated and perforated. The only direct opening to the outside is the main entrance that gives access to the sales area. By changing the interior light and the symbol of pharmacy, the building gains dynamic, allowing the image variation from day to night. The store not only sells medicines but also has its own laboratory for compounding pharmacy, creating a one-stop shop for shoppers.

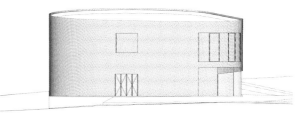

East elevation　东立面图

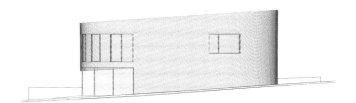

North elevation　北立面图

1. The big logo with the green lighting
2. East exterior view
3. Entrance

1. 散发出绿光的大型标识
2. 建筑东侧外观
3. 入口

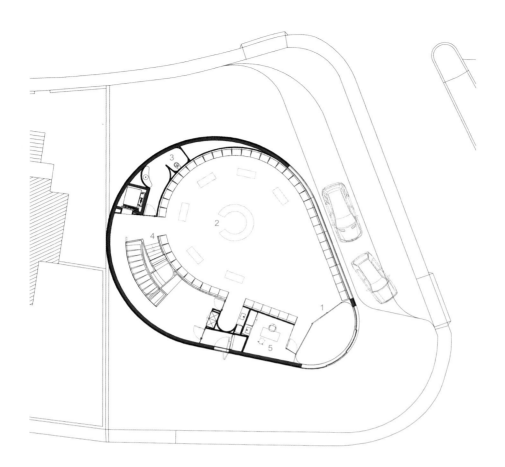

Ground floor plan

1. Entrance
2. Showcase
3. Bathroom
4. Staircase
5. Office

一层平面图

1. 入口
2. 展示区
3. 洗手间
4. 楼梯
5. 办公区

1. Preparing area
2. Counter
3. Integral interior view
4. Waiting area

1. 配药区
2. 柜台
3. 综合室内设计
4. 等候区

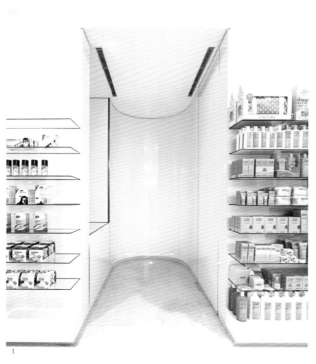

药店位于葡萄牙中北部的雷阿尔城。雷阿尔城位于城市的外围，其周边环境没有固定统一的模式。由于没有外观结构的参考，药店的建筑设计抽象独特，外部几乎没有门窗开口。两层高的椭圆造型全部采用波纹镂空铝板包裹。建筑与外界唯一直接相连的地方就是通往售货区的主入口。室内灯光和药房标志的变幻让建筑显得动感十足，日夜变化。药店不仅销售药品，还有自己的复方药实验室，为消费者提供了一站式服务。

1. Showcase
2-4. Interior details

1. 展示区
2~4. 室内细部设计

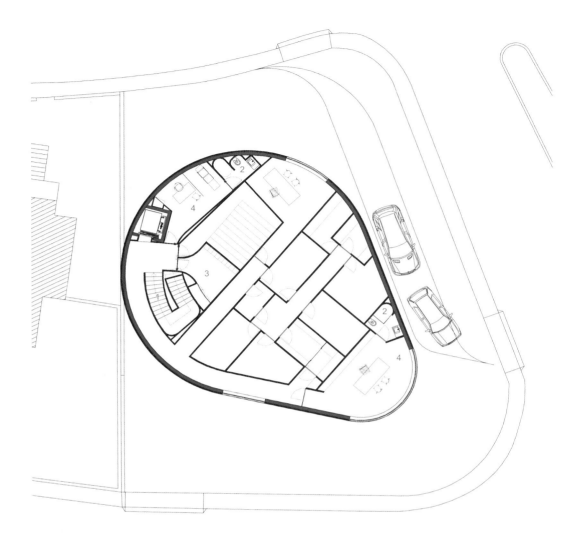

First floor plan

1. Staircase
2. Bathroom
3. Drug dispensary
4. Offices

二层平面图

1. 楼梯
2. 洗手间
3. 配药室
4. 办公室

1. Meeting room
2. Drug dispensary
3. Staircase
4. Corridor

1. 会议室
2. 配药室
3. 楼梯
4. 走廊

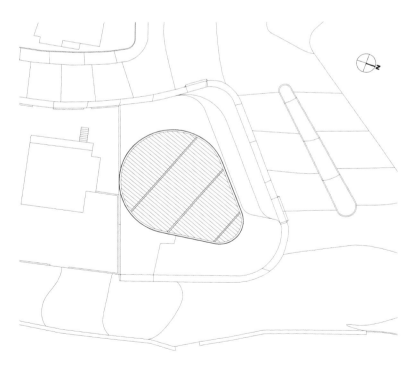

Site plan 总平面图

CHAPTER 2 **Cases Study** | Detached Commercial Architecture

Completion date: 2010
Location: Los Angeles, USA
Designer: M Sc.Ton Smulders
Architect: Patrick Kinmonth, Space Architects
Photographer: Courtesy Missoni Archives
Construction area: 700sqm

Missoni Store 米索尼服装店

Key words: Architectural Translation with Simple Lines and Figurative Symbols
The exterior lines are simple and direct with pure and bright white colour. The corner location helps the project to stand out. The facade is covered with overlapped white brushed aluminium strips, forming a undulating flow effect to symbolise the brand's knitwear products. At night, light filters through the gaps between the bands, providing a dynamic effect for the building.

关键词：线条简约、比喻具象的"建筑翻译语言"
该项目外部线条简单、直接、色彩明亮、纯净，由于所处位置为街区的转角处，项目更为出挑。建筑外立面由纯白磨砂铝条横向交叠，形成波浪形的流动效果，蕴含着该项目纺织品牌的喻意。夜晚时，灯光从铝条间的缝隙透过，使得建筑更具动感。

Design Points:

The Missoni flagship store in Los Angeles looks like a curious inhabitable object at the corner of the famous shopping routes of Beverly Hills, Rodeo Drive and Santa Monica Boulevard. Marzorati Ronchetti has produced and installed the unusual, expressive façade solution, which in the intentions of the designers symbolises and narrates the values of the productive tradition of Missoni.

In effect, the complex pattern of strips of aluminium, with an irregular undulating shape, positioned horizontally, reminds people of the weave of woollen yarns in the famous knitwear products by Missoni, though in this sort of "architectural translation" the threads are purged of their refined chromatic mélange and become an iconic pattern in absolute white. The powder-coated aluminium strips (10cm in height, 8mm thick, in lengths varying from 50cm to 7m) are overlapped and supported, imitating the weave of yarns on a loom, on a concealed steel structure.

The regular parallelepiped of the store is completely covered by the resulting weave, which is also clearly visible from inside. The architectural skin is interrupted and sculptured by the opening for the recessed entrance door and those of the shop windows, which unite the retail space with the scene on the street.

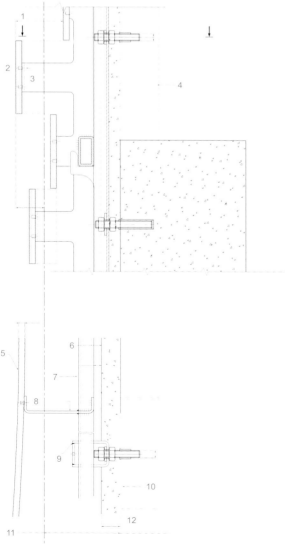

Inside, daylight filters through the façade pattern, generating striking effects of light and shadow, while at night the gaps between the horizontal bands come alive with light from the interior, making the monolith into a porous, sparkling, light and lively form.

Detail drawing	细部图
1. 62.5 maximum gap	1. 最大间隙62.5
2. Aluminium plate 100*8 Colour T.B.D.	2. 铝板100X8 T.B.D.色
3. Screw M4 countersink head Inox	3. Inox不锈钢埋头孔M4螺栓
4. Steel profile by other	4. 钢材剖面
5. Aluminium plate	5. 铝板
6. Stucco	6. 灰泥
7. AISI304 frame Pipe 40*20*3 Colour T.B.D.	7. AISI304不锈钢框架 40X20X3管 T.B.D.色
8. Sheet AISI304 THK.3 Colour T.B.D.	8. AISI304不锈钢板，厚度为3 T.B.D.色
9. Secondary structure AISI304 plate THK.3	9. 二级结构 AISI304不锈钢板，厚度为3
10. Steel profile by other	10. 钢材剖面
11. Working point	11. 工作点
12. Upper concrete beam line	12. 上方混凝土梁线

1. Exterior view
2. The glass display window

1. 建筑外观
2. 玻璃橱窗

CHAPTER 2 **Cases Study** | Detached Commercial Architecture

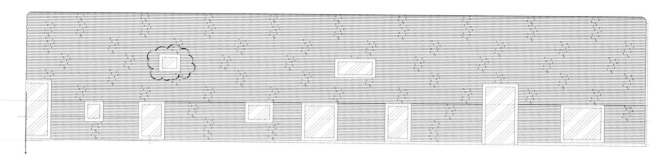

Elevation 立面图

洛杉矶米索尼旗舰店坐落在比弗利山庄、罗迪欧大道和圣塔莫尼卡大街三条著名商业街的转角，看起来像一座新奇的住宅。马卓拉蒂·罗克迪亲自打造了这一奇特、夸张的外立面结构，旨在表现米索尼品牌多样创新的传统。

事实上，由不规则波纹的水平排列所形成的复杂的铝条造型让人想起了米索尼品牌著名的编织类产品上的粗纺毛纱。在这个"建筑翻译"中，纱线失去了亮丽的色彩，以纯白色来凸显图案造型。带有粉末涂层的铝条（10厘米高、8毫米厚、长度在50厘米到7厘米之间）相互叠加和支撑，在钢铁结构上模仿了纱线在织布机上的情景。

店铺规则的平行六面体结构全部由编织图案覆盖，这种结构即使在建筑内部也清晰可见。嵌入式大门和店铺橱窗打断了连续的建筑外壳，将店铺内部的零售空间与街景联系起来。

阳光透过建筑立面的图案进入室内，形成了条纹样的光影。夜晚，室内的光线让水平条带之间的空隙变得鲜活，使建筑显得透气、闪耀、明亮而又活泼。

Detail drawing

1. Aluminium plate 100*8
   Colour T.B.D.
2. Screw M8 countersink head inox
3. Aluminium plate 100*8
   Colour T.B.D.
4. Stucco
5. Silicone
6. Lamp rail support
7. Button holes
8. Hole 25mm on each side
9. Silicone
10. Enclosure sheet for lamp box
11. Lamp box
12. Visarm 8+8
13. Working point
14. Upper concrete beam line
15. Conduits by others
16. Plasterboard
17. Silicon bubble
18. Sprinkler
19. Adhesive plastic film for protection

细部图

1. 100X8铝板
   T.B.D.色
2. Inox不锈钢埋头孔M8螺栓
3. 100X8铝板
   T.B.D.色
4. 灰泥
5. 硅树胶
6. 灯轨支架
7. 扣眼
8. 两侧各有一个25mm孔
9. 硅树胶
10. 灯箱外壳板
11. 灯箱
12. Visarm玻璃板8+8
13. 工作点
14. 上方混凝土梁线
15. 导管
16. 石膏板
17. 硅胶泡
18. 洒水器
19. 塑料保护膜

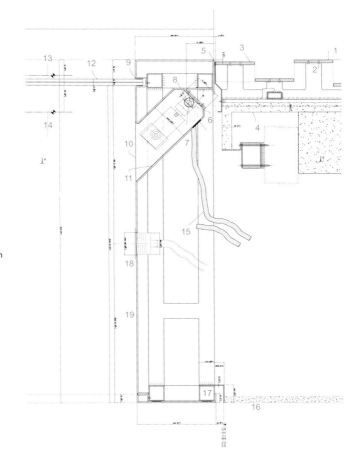

1. The exterior view in night
2. Shopfront
3,4. Façade detail

1. 建筑外观夜景
2. 店面
3、4. 外立面细部设计

| Completion date: 2010 | Designer: Enrique Ramirez, Mark Chapman | Photographer: Panos Kokkinias |
| Location: Athens, Greece | | Construction area: 600sqm |

Placebo Pharmacy 安慰剂药店

Key words: Solidness and Transparency Consisted by Different Materials and Structure of the Building Envelop
The designer reforms the octagonal shape of the existing structure into a cylinder in order to create a spiral which seeks to converse with the rapid motion on the street where the building stands. The new envelop uses two different materials and structures: the panels of the ground floor are perforated using Braille, while the upper floor is enclosed with wood gratting. These two envelop structures overlaps and climbs up spirally, achieving changes of solidness and transparency with light changes.

主题词：材质和结构不同的表皮构成的虚实变幻
该项目的设计师将原有八角形的建筑外观改造为圆柱体，并形成了螺旋向上的趋势，这正与项目所处位置川流不息的车流相呼应。建筑的新表皮采用了两种不同的材质和结构：地面层采用镂空的、雕刻有盲文的白色面板；二层部分采用木格栅。这两种看似不同的表皮结构相互交叠，盘旋上升，使建筑在光影交替中实现了虚实变幻。

Design Points:

The design process for this large supralocal pharmacy forced the designers to shift their viewpoint and come up with a virtual building – a placebo pharmacy. The octagonal shape of the existing structure was reformed into a cylinder in order to create a spiral which seeks to converse with the rapid motion on Vouliagmenis Avenue, the urban artery on which the building stands. The panels of the façade are perforated using Braille, which both alludes to the system's use on pharmaceutical packaging and boosts visibility by allowing the light to find its way into the interior. The new façade also protects the interior while acting as a lure for passers-by. The changes of the natural light coming through the Braille holes inside the store are inevitably playful according to the sun route.

Inside, the product display mirrors the circular frontage, while a ramp up to the upper level extends the dynamism of the exterior spiral into the interior space.

The concept was to create a pharmacy that is typical "a pharmacy", but rather a unique environment that inspires. It represents a pharmacy for healthy people who believe in prevention opposed to the aftermath

process of curing and treating.

The shelving units act simultaneously as dividers for the open floor plan, naturally creating aisles. The form of the "furniture", (shelving units), resemble the shape of a leaf or petal made of white lacquered MDF, which are internally illuminated. The idea behind the way the products are presented on the shelving units follows the norms of a museum. Each item is exhibited and considered "sacred".

The "real" pharmacy lies behind a stainless steel counter where the drugs are located within a wall of stainless steel drawers. The materiality the designer employs for this project is chosen deliberately to provoke a laboratory ambiance, which in essence generates a feel for an existing natural environment. These all contributed to the idea that

1. The entrance
2. West exterior view
3. East exterior view

1. 入口
2. 建筑西侧外观
3. 建筑东侧外观

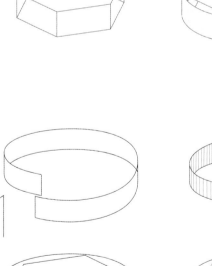
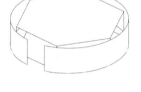
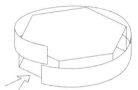

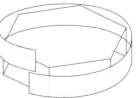

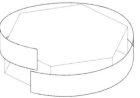
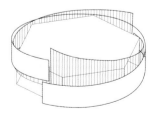

1. Ramp and the office area
2. The ramp from ground floor to 1st floor
3,4. Radial pattern showcase

1. 坡道和办公区
2. 坡道从一楼延伸到二楼
3、4. 辐射型展示区

Concept diagram
设计概念图解

nature co-exists with technology innovation to create pharmaceutical and cosmetic products.

The floor fabricated by white epoxy resin initiates a uniformity of the layout, whereby on the ground floor, cut-outs in natural forms are curved inside the wall in order to illuminate the pharmacy with the use of RGB LED lighting.

The Pharmacy is composed of two floors, the ground floor and the upper mezzanine level. The ground floor exists as the main shopping area of the pharmacy, while the upper mezzanine floor, consists of additional office space as well as a communal, multi-functonal area, where events can be held. The partitions separating the different rooms for different uses are made of green transparent glass. One can view the below level by looking down from the existing exterior corridor onto the monumental ramp leading to the ground level.

The floor plan of the pharmacy follows a radial pattern with the main cashiers desk acting as the focal point. The shelving units are arranged in such a manner that they seem to fan out from the focal point presenting the cashier with the ability to view the whole pharmacy from its central position. The drug dispensary, the preparational areas and the toilets are also arranged according to this radial pattern. The pattern provides the space with a natural flow of circulation while simultaneously allowing light to move towards the centre of the floor plan throughout the entire day.

1. The landscape bamboo between interior and exterior
2. The ramp

1. 室内外夹层中的景观竹子
2. 坡道

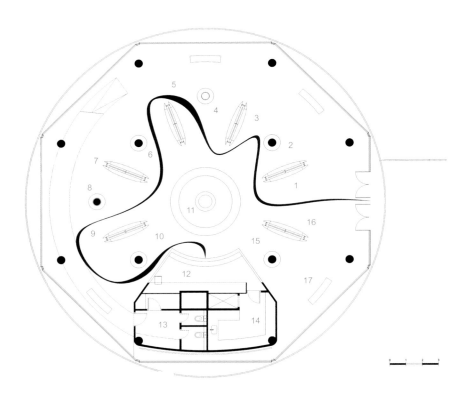

Ground floor plan

1. Dental products
2. Optican
3. Men cosmetics
4. Diet
5. Smokers
6. Electronics
7. Mother and baby
8. Vitamins
9. Orthopedics
10. Winter goods
11. Make-up cosmetics
12. Drugs dispensary / check out
13. WC
14. Drugs preparation
15. Summer goods
16. Books health and diet
17. Supermarket herbs and food

一层平面图

1. 牙科产品
2. 眼药
3. 男士化妆品
4. 减肥药
5. 吸烟用药
6. 电子仪器
7. 母婴药品
8. 维生素
9. 矫形用品
10. 保暖产品
11. 化妆品
12. 配药房/收银台
13. 洗手间
14. 备药室
15. 消暑产品
16. 健康与饮食图书
17. 自选药草和食品

这座大型药房的设计流程改变了设计师的视角，最终建成了一座虚拟建筑——安慰剂药店。原有结构的八角形经过改造成为了圆柱形，其所形成的螺旋结构与建筑所在的乌里格莫尼斯大道——一条城市主干道——相互呼应。建筑外立面的金属板以盲文形成了镂空结构，既暗示了药品包装上的盲文，又让室外的光线有效进入室内。新建的外立面在保护室内环境的同时，还起到了吸引路人的作用。自然光透过盲文孔渗透到药店内部，并且随着太阳的轨迹不断变换。

药店内部的商品陈列与建筑外观的圆形结构遥相呼应。通往二楼的坡道将室外动感的螺旋结构延伸到了室内空间。

项目的设计概念不是建造传统意义上的"药店"，而是打造一个能够鼓舞人心的独特环境。它代表着一种供健康人使用的药剂，其目的不在生病以后治疗，而是在于预防。

货架同时也是分隔开放式空间的隔断，在药店内部自然地形成了走道。货架的造型模仿了叶子和花瓣，采用白色烤漆中纤板制造，从内部进行照明。商品在货架上的陈列方式效仿了博物馆的陈列方式。每样商品都被精心摆放出来，充满了"神圣感"。

"真正"的药房在一个不锈钢柜台的后面，药品被放置在墙上的不锈钢抽屉里。设计师所选择的的材料刻意模仿了实验室的氛围，令人有置身自然的感觉。所有设计都直指一个概念——自然与技术创新共同创造了药品和化妆品。

白色环氧树脂地面与整体布局相一致，一楼墙壁上被切割出若干个自然图形凹孔，用以设置红绿蓝三色LED灯为药店照明。

药店由两层楼组成——一楼和上层的夹层楼。一楼是主要的售货空间，而夹层楼则属于附加的办公、公共及多功能空间，可以举办各种活动。分隔各个房间的隔断墙采用了绿色透明玻璃。人们可以从原有的室外走廊透过独特的坡道看到下层的景象。

药店的楼面布局呈放射形，以收银台为中心。货架呈扇形从中心点向外辐射。药房、过渡区和洗手间同样采用了这种放射形布局。这种布局为空间提供了自然流畅的交通，同时也让日光能够整天都照射到建筑内部。

3D construction sketch
三维结构草图

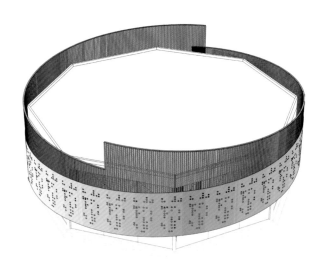

1. Drug dispensary
2. Interior view of 1st floor
3. Braille perforated façade panel
4. Meeting room

1. 配药室
2. 二楼内景
3. 盲文镂空面板
4. 会议室

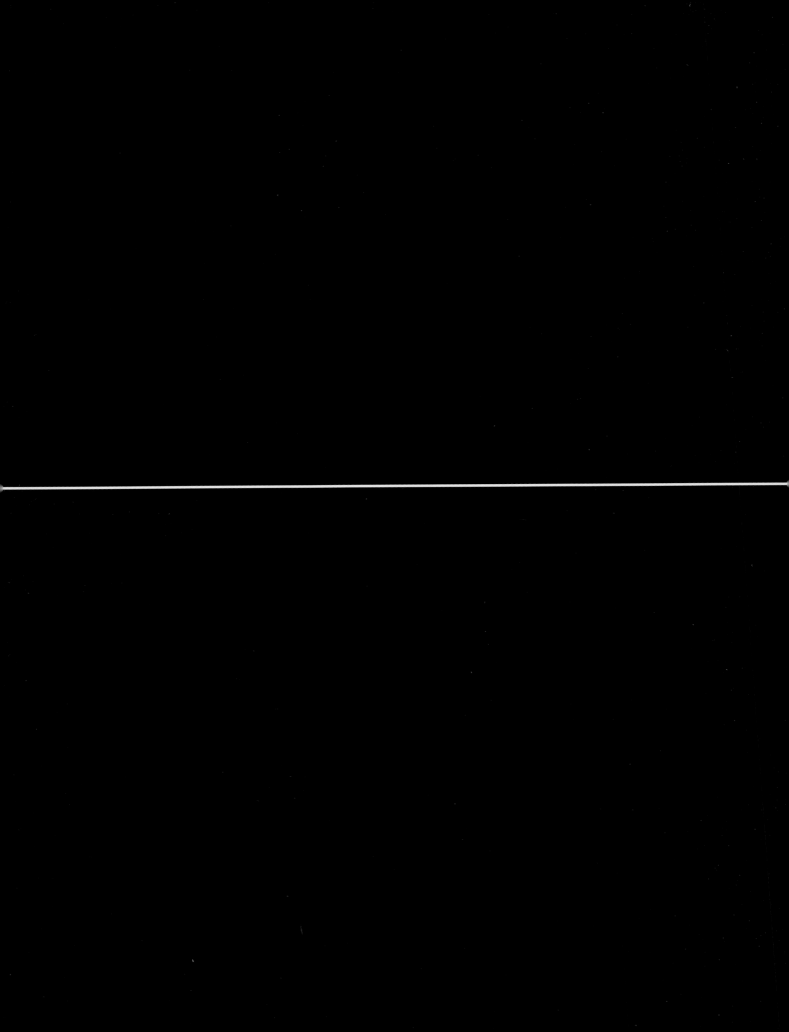

When the environment and its colours are perceived, the brain processes and judges what it perceives on an objective and subjective basis. Psychological influence, communication, information, and effects on the psyche are aspects of our perceptual judgment processes. Hence, the goals of colour design in an architectural space are not relegated to decoration alone. Empirical observations and scientific studies have proven that human-environment-reaction in the architectural environment is to a large percentage based on the sensory perception of colour.

当人眼看到环境及其色彩后,大脑根据主客观对其进行处理和判断。心理影响、交流、信息和精神都是我们的感知判断流程。因此,建筑空间色彩设计的目标不仅局限于装饰。观察和科学研究证明,建筑环境中的人与环境之间的相互反映大部分都以色彩的感知为基础。

ARCHITECTURE AND COLOURS

建筑与色彩

3.1 Concept

Light projects to the architecture's exterior wall, then the wall materials absorb certain amount of light and reflect it into human's eyes. The visual effect that it generated is architectural colour.

3.2 Three Properties of Colours

3.2.1 Luminance and Brightness

Luminance is a photometric measure of the luminous intensity per unit area of light travelling in a given direction.

Brightness is a judgmental measure of comparison between the luminance in a certain surface and the luminance in a white surface.

3.1 概念

建筑色彩就是由光投射在建筑物外墙上,外墙材质吸收一定光,又把其余的光反射到人眼所产生的视觉效果。

3.2 色彩的三属性

3.2.1 亮度、明度

亮度是表示从某一领域发出的光能看到较多还是较少的感觉的概念。

明度是将某一面的亮度和与其同样照射的白色面的亮度进行比较而判断的量。

1. The colourful metal plates wrapping the façade
2. The interlaced black and yellow

1. 包裹于建筑外立面色彩各异的金属板
2. 黑色与黄色的交错搭配

3.2.2 Hue

Hue is one of the main properties of a colour, defined technically as the degree to which a stimulus can be described as similar to or different from stimuli that are described as red, green, blue, and yellow.

3.2.3 Colourfulness and Saturation

Colourfulness is the degree of difference between a colour and gray. It is the absolute amount of hue.

Saturation is the purity of a colour. The level of saturation is values by the proportion that the pure colour with the same name takes up in a certain colour. High proportion means high saturation and low proportion means low saturation.

3.3 Cognition of Colours

3.3.1 Warm Colours and Cool Colours

Warm colours are often said to be hues from red through yellow, browns and tans included, which provide warm feelings. Cool colours are often said to be the hues from blue green through blue violet, most grays included, which provide cool feelings. When the saturation increases, these influences will increase too.

3.3.2 Advancing Colours and Receding Colours

Despite the same distance, some colours seem near while some seem far. The former colours are advancing colours while the latter receding colours. The most advancing colour is red, followed by orange, yellow, purple red and blue.

3.3.3 Expansive Colours and Contractive Colours

Despite the same sizes, some colours seem to be large while some seem to be small. The former colours are expansive colours while the latter contractive colours. The most expansive colour is yellow, followed by green, red and blue.

3.2.2. 色相

色相是色彩所呈现的质的面貌，是色彩彼此之间相互区别的标志。色相是色彩的首要特征，是区别各种不同色彩的最准确的标准。

3.2.3 视彩度、彩度

视彩度，是表示某一领域色彩的色相的量的多少的量，即相当于色相的绝对量。

彩度，是指色彩的纯度，通常以某彩色的同色名纯色所占的比例，来分辨彩度的高低，纯色比例高为彩度高，纯色比例低为彩度低。

3.3 色彩的认知

3.3.1 暖色、冷色

顾名思义，暖色是指红、黄、橙等看上去给人温暖感觉的色彩；相反为冷色，冷色就是指看上去给人感觉寒冷的色彩，比如蓝色或与其相近的色彩。如果彩度增加，这些影响亦会增加。

3.3.2 前进色、后退色

尽管距离相同，但是有的色彩看上去近在眼前，有的颜色看上去却比较遥远，前者称为前进色，后者称为后退色。最为前进的色彩是红色，随后为橙色、黄色、紫红色、蓝色。

3.3.3 膨胀色、收缩色

尽管大小相同，但是有的色彩看上去比较大，有的色彩看上去比较小，前者称为膨胀色，后者称为收缩色。色彩越为明亮越有膨胀感，最有膨胀感的是黄色，其次是绿色、红色和蓝色。

3.3.4 Heavy Colours and Light Colours

Compared with objects with bright colours, objects with dark colours seem to be heavier. The weight of colours reduces in the order of black, red, blue, purple, violet, green, orange, yellow and white.

3.3.5 Association and Symbolism

Different colours will cause different psychological feelings. Red suggests passion and boldness; blue suggests sky and ocean; purple suggests romance…. These colours are associated to not only specific objects, but also abstract conceptions and mental status. This abstract association is called colour symbolism.

3.3.6 Colour Emotions

Colour emotions are a series of colour psychological reactions generated by human brain when optical information with colours of different wave lengths act on eyes and be passed to the brain. Slight changes in saturation or opacity will create totally different feelings.

3.3.4 重色、轻色

色彩较暗的物体与色彩较亮的物体相比，看上去更有重量感。彩度的影响根据色相而不同，颜色的重量感以黑、红、蓝、紫、堇、绿、橙、黄、白的顺序减弱。

3.3.5 联想与象征

人们看到不同的颜色会产生不同的心理感受，红色使人想到热情、奔放，蓝色让人想到天空、海洋，紫色有种浪漫的情感……这些让人产生联想的不仅仅是具体的事物，还有抽象的东西和精神状态，这种抽象的联想就是色彩的象征性。

3.3.6 色彩感情

色彩感情指不同波长色彩的光信息作用于人的视觉器官，通过视觉神经传入大脑后，经过思维而形成一系列的色彩心理反应。每种色彩在饱和度、透明度上略微变化就会产生不同的感觉。

3. Sketch of Ferrari Factory Store

3. 法拉利工厂店手绘图

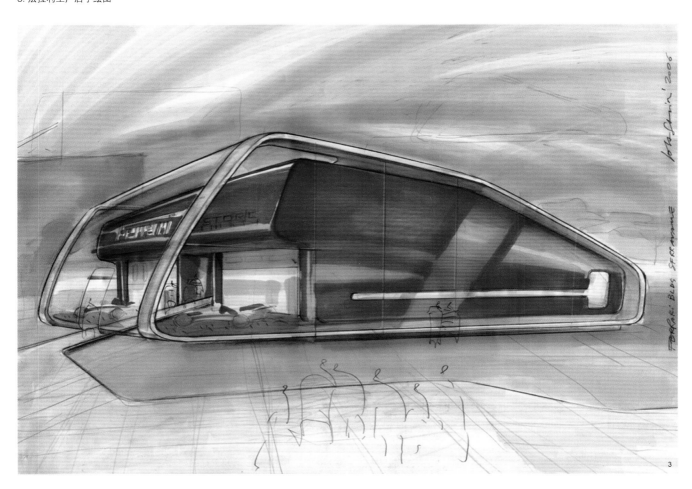

Colour Properties 色彩属性	Colours 色彩	Colour Emotions 色彩感情
色相 Hues	红色 Red	激情、愤怒、欢喜、积极、兴奋 Passion, Anger, Happiness, Positivity, Excitement
	橙色 Orange	喜悦、欢闹、活泼、健康 Joy, Playfulness, Vividness, Health
	黄色 Yellow	快活、明朗、愉快、积极、天气 Liveliness, Brightness, Pleasance, Positivity, Weather
	绿色 Green	轻松、舒适、平静、年轻 Easiness, Comfort, Calmness, Youth
	紫色 Purple	严肃、神秘、不安、温柔 Seriousness, Mystery, Uneasiness, Gentleness
	青绿色 Turquoise	安息、清凉、忧郁 Rest, Coolness, Melancholy
	蓝色 Blue	深沉、寂寞、悲哀、深远、镇静 Depth, Loneliness, Sadness, Far, Calmness
	蓝紫色 Violet	神秘、崇高、孤独 Mystery, Highness, Loneliness
明度 Brightness	白色 White	纯粹、清爽 Purity, Coolness
	灰色 Grey	沉着、压抑 Serenity, Depression
	黑色 Black	阴暗、不安、庄严 Darkness, Uneasiness, Solemnity
彩度 Saturation	红色 Red	热烈、激烈、热情 Hotness, Intensity, Passion
	粉红色 Pink	可爱、温柔 Loveliness, Gentleness
	暗褐色 Puce	沉着 Serenity

3.3.7 Memory Colours

In people practice, they form deep memories of certain colours. Therefore they maintain some regular recognitions for these colours and form inherent habits. These colours are called memory colours. With same saturation, compared to cool colours, warm colours are easy to remember.

3.4 Preparations for Architectural Colour Design

- Design Drawings
Architectural design scheme
Architectural drawings, including maps, floor plans, elevations, sections, expansion drawings, door and window listings, decoration listings, etc.

- Colour Codes
International standard colour code
Standard colour code for architecture
Standard colour code for paint
Material colour samples

3.3.7 记忆色

人们在长期实践中对某些颜色的认识形成了深刻的记忆，因此对这些颜色的认识有一定的规律并形成固有的习惯，这类颜色就称为记忆色。彩度相同时，与冷色系的色彩相比，暖色系的色彩易于记忆。

3.4 建筑色彩设计的前期准备

- 设计图纸类
建筑设计方案书
建筑设计图（配地图、平面图、立面图、剖面图、展开图、门窗列表、装饰列表等）
- 色标类
国际标准色标
建筑用标准色标
涂料用标准色标样本书
材料颜色样本

4. Colours are used in the undersides of the panels
5. The brand colour is presented in the façade

4. 将色彩用于各层板面下缘
5. 品牌色彩在外立面的设计表现

3.5 Considerations about Architectural Colour Design

• It should correspond to architecture's function; interior colour design should correspond to space requirements.
• It should coordinate with the surroundings (e.g., urban plan, adjacent landscape, district requirements); exterior colour design should consider the surrounding colour scheme, including colours of natural scene.
• It should coordinate with local cultures.
• As a background, it should highlight people and objects.
• It should be accepted by most people.
• It should consider local climate. Though the exterior wall uses the same materials and colours, when the uneven surface is obvious, the chromaticity will show great differences in different illumination levels. The colours will fade gradually from sunny day, cloudy day and rainy day to foggy day.

3.6 Colour Categories of Architectural Components

The colours of architectural components can be divided into three categories: basic colours, matching colours and accent colours. Basic colours are used in background and occupy a relatively large area. They decide the overall atmosphere. Matching colours are the colours of graphics and their area is second to that of basic colours. They play the role of performance. Accent colours utilise small areas to highlight the overall colour impression.

3.7 Reinforcement and Restraint of Architectural Colours

Consider to reinforce or restrain the colour application according to space composition, materials and size of the building.

Main parts, structures with unique forms and bright parts should be reinforced. Strong colours, white and black should be used.

Reversely, large areas, areas with basic colours, dim parts and sunken

3.5 建筑色彩应该注意的几点

• 应该符合建筑功能，室内色彩设计应符合空间需求
• 应该与周围的环境（城市规划、周边景观、街区要求）协调一致，在进行建筑物的外部色彩设计时要考虑到包括自然景物在内的周边环境色彩
• 应该和当地的文化协调一致
• 应该作为突出人与物的背景而存在
• 应该为大多数人所接受
• 应该考虑天气的影响。即便是相同材质、色彩统一的外墙面，如果凹凸面较为明显，当光照强度不同时，在色度上也会表现出较大的差异。按照直射、阴天、雨天、雾天的顺序，色彩的色泽会逐渐变淡

3.6 建筑构成要素的色彩分类

建筑的构成要素的色彩根据面积可以分为基调色、配合色和突出色。基调色是背景部分的色彩，占据较大面积，起到决定整体氛围的作用。配合色是图形部分的颜色，所占面积仅次于基调色，具有表现特征的作用。突出色具有通过强调小面积部分突出色彩整体印象的作用。

3.7 建筑色彩的强化与抑制

色彩的应用需根据建筑的空间构成、材料、大小等考虑强化或者抑制。

重要的部分、外形独特的结构、光线明亮的部位等是应该强化的地方，应使用强色彩和黑色、白色。

相反，面积较大的部分、基调色位置、光线较弱的部位、凹陷的部位等应该是抑制的部分。多使用弱色彩，即除了白色、黑色以外的无彩色、超灰色等或低彩度的暖色。

parts should be restrained. Weak colours, i.e. neutral colours except for white and black, gray or warm colours with low saturation should be used.

3.8 Construction Materials' Influences on Colour Selection

Consider the different construction materials' surface characteristics in material selection. For example, colours of expressive components with textures and shading patterns will be more attractive when reinforced; glossy parts are acceptable even when strong colours are used.

3.9 Construction Materials and Colours

3.9.1 Wood
Wood is perishable and inflammable. However, with rigid working condition and good preservation, it is also durable. Wood is easy for processing and widely used. Compared with metal and plastic, wood feels warm and natural. Generally, wood has warm colours such as yellow, brown and red.

3.9.2 Stone
Stone possesses high strength and durability. As it is composited with one or several kinds of minerals, its colour is influenced by colours of minerals with high proportions, adjacent mineral's colours and minerals' grain sizes. With beautiful gloss, polished natural stone is rich in kinds and has unique colours. Mostly, stone evokes heavy and cold feelings.

3.9.3 Concrete
Concrete is a composite material composed of coarse granular material embedded in a hard matrix of material that fills the space among the aggregate particles and glues them together. It is mainly in dark grey colour.

3.8 建筑材料对色彩选择的影响

由于建筑材料的选择不同，在选择材料时应考虑到材料的表面特性。带有纹理、底纹图案等表现强烈部位的色彩经过强调后更加醒目，富有光泽的部位即便使用强色彩也易于接受。

3.9.4 Glass
Glass is incorruptible and possesses a high translucency. Its translucency can be altered through surface finish. By adding trace amounts of metal, glass can be coloured to control light absorption. In some cases, glass may be the decisive factor of the building's first impression. Not only different types of glass can lead to different surface colours, interior landscape, lighting and curtains can also influence the integral colour of glass.

3.9.5 Paint
The colour of paint is decided by the secondary coating. The paint for secondary coating is mixed with several primary colours. The colouring material as colour source is called pigment.

3.9.6 Metal
Metal materials include plain steel, stainless steel, titanium, cast iron, etc. Metal façades are colourful with unique metal gloss, rich in textures.

6. ETECH Shop Linz is decorated in orange in the whole
7. The attractive neon-yellow of Bershka Shibuya
8. The pure white façade looks simple yet elegant

6. 整体运用橘色装点的ETECH林茨店
7. 波丝卡涩谷店引人注目的一抹霓虹黄
8. 素雅的白色立面简洁、大方

3.9 建筑材料与色彩

3.9.1 木材
木材具有易腐、易燃的缺点，但经过严格的使用条件和良好的保存处理，也能具有长期的耐久性。木材易于加工，用途广泛，与金属和塑料等工业材料相比，木材感觉温暖、自然，色彩多以黄、棕、红的暖色为主。

3.9.2 石材
石材强度高并具有耐久性，一般由一种或多种矿物质组成，其色彩受到含量比较高的矿物质自身色彩、相邻矿物质的色彩及矿物质的粒径的影响。自然石材经过打磨而具有美丽光泽，种类丰富，色调独特，多给人以厚重和寒冷的感觉。

3.9.3 混凝土
混凝土是在水泥和沙子中以适当的比例混合沙砾、碎石等骨材，用水搅拌、硬化而成。可以使用模板制作成多种形状，以暗灰色为主。

3.9.4 玻璃
玻璃不易腐蚀，具有高透光性，通过表面加工可以为透视性增加变化。在原料中添加微量的金属类可以进行着色，控制光的吸收率。玻璃有时会成为决定外观印象的重要因素，不仅玻璃的种类可以导致玻璃本身和表面的色彩不同，而且设置于室内的景观、灯光、窗帘等物品都会影响到整体的外观色彩。

3.9.5 涂料
涂料的色彩由二次涂抹决定，二次涂抹的涂料分别有几种原色体系，使其原色混合进行色彩调和，作为这种色彩的源泉的着色材料称为颜料。

3.9.6 金属类
金属类材料中有普通钢、不锈钢、钛、铸铁等。金属类外立面色彩丰富，多带有金属特有的光泽，有质感。

Completion date: **2010**	Designer: **Castel Veciana Arquitectura, KEY OPERATION INC./ARCHTECTS.**	Photographer: **Hattori Studio**
Location: **Tokyo, Japan**		Construction area: **1,413.67sqm**

Bershka Shibuya 波丝卡涩谷店

Key words: Using Neon-yellow Colour to Highlight Building's Position
Located in fashion district of Shibuya, the store is positioned on young people who pursuit fashion. In order to achieve high visible effect, the designer chooses a direct visual impact to attract attentions. The floor, walls, stairs and pillars of the staircase are decorated with eye-catching neon-yellow. Seen through the glass curtain wall, the pop neon-yellow colour is dazzling even at night.

主题词：一抹霓虹黄凸显建筑位置
该项目地处涩谷时尚区，店铺销售群体定位于追求时尚的青年男女，为了使整个建筑出挑，设计师选择用最为直接的视觉冲击来夺人眼球，在楼梯间的地面、墙面、楼梯本身乃至楼梯支柱等处都使用了最为抢眼的霓虹黄。透过外立面的玻璃幕墙，跳跃的霓虹黄色格外耀眼，即使夜幕降临，也未逊色丝毫。

Design Points:

Bershka is a casual fashion brand originating in Spain close in 1998, with a customer target of fashionable young women and men in their teens and early 20's. Japan is country No. 50 for Bershka, and Shibuya was chosen as the beginning point for the brand in Japan, the first of several projects to be rolled out this year. The Shibuya store is an outstanding example of a Bershka flagship store. Located at the foot of Spain-zaka in one of the best locations in the highly competitive fashion district of Shibuya, the Bershka store opens on 4 floors in the ZERO GATE building.

The façade – The visual impact of the staircase structure in pop neon-yellow colour unites all the floors, as a sign of the Bershka identity that can be perceived from the exterior. The backlit logo sign runs vertically, making a clear statement.

Interior – The interior feel of each floor corresponds to the products on each floor. White-toned or dark-toned spaces, walls and ceilings with signage graphics give dynamism to the complex. Store furniture is simple and functional. The electric escalators and shopping spaces are divided by a wall in which LED lighting runs up and down, inviting customers to move to the upper floors, in addition to being an illumination spectacle.

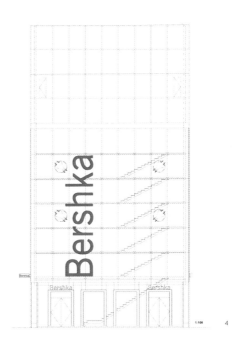

1. Exterior view
2. Shopfront design
3. Exterior view in night
4. Shopfront elevation

1. 建筑外观
2. 店面设计
3. 建筑外观夜景
4. 店面立面

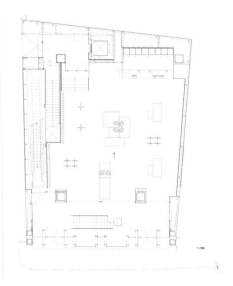 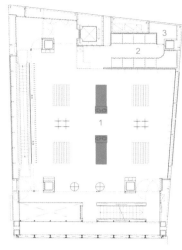

波丝卡是一个西班牙休闲流行品牌,创立于1998年,目标消费人群为追求时尚的青少年。日本是波丝卡进军的第50个国家,涩谷成为该品牌进军日本的先锋城市,因此在这一年被作为首选项目。涩谷店是波丝卡旗舰店在日本的一个杰出代表,坐落于竞争激烈的涩谷时尚区的Spain-zaka大街,是一个四层高的独立店铺。

外观。楼梯结构使用的是富有视觉冲击力的流行的霓虹黄色,作为BERSHKA的标志,可以从外面就被看到。背光标志牌垂直显示,十分突出。

室内。室内设计每层的基调与相对应的产品呼应。产品标示与白色调与深色调的空间,墙壁和天花板给复制的空间带来一种活力感。店铺家居设计简单,功能性强。

电动扶梯和购物空间被一面墙分开,墙上的LED灯光上下转换,除了作为一个照明景象,还可以吸引顾客移动到更高的楼层。

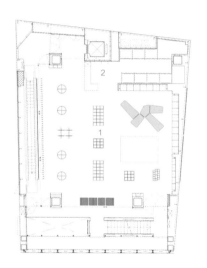 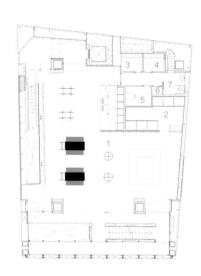

Floor plans
平面图

1. Man	1. 男装区
2. Fitting room	2. 更衣室
3. Stock 1	3. 1号仓库
4. Stock 2	4. 2号仓库
5. Office	5. 办公区
6. Stock 3	6. 3号仓库
7. Restroom	7. 洗手间

1. Ground floor interior
2. First floor interior
3. Staircase
4. The showcase of the ground floor

1. 一层室内设计
2. 二层室内设计
3. 楼梯
4. 一层展示区

1. The 2nd floor interior
2. The fitting room of 1st floor
3,4. The showcase of 2nd floor
5. The showcase of 1st floor

1. 三层室内设计
2. 一层更衣室
3、4. 三层展示区
5. 二层展示区

Completion date: 2011
Location: Linz, Austria
Designer: kleboth lindinger dollnig
Photographer: Nik Fleischmann
Construction area: 1,250sqm

ETECH Shop Linz ETECH 林茨店

Key words: Achieving High Recognition
In order to stand out from its surrounding large buildings and vast parking lots, the design selects an exclusive sign, a unique shape and a highly visible orange colour. The stacked wooden blocks distinct the building from its surrounding standard-looking buildings and clarify the building's interior functional zones. The combination of warm orange and the exclusive sign is highly recognisable.

主题词：实现建筑的高识别度
为了使被大型建筑物和广阔的停车场包围下的建筑脱颖而出，项目选择采用专属的招牌设计，特异的建筑造型和穿透力极强的单一橘色。积木式的建筑造型将项目与其周边中规中矩的建筑相区分，也使得其室内实际使用空间更加清晰；暖色系的橘色明视度高，搭配专属的店面招牌，辨识度极高。

Design Points:

Amidst the competition of shops and shopping centres to attract attention of passers-by, a key challenge is to leave a lasting impression on potential customers. This is particularly true if David is up against Goliath: a small but well-stocked local retail chain against the global chain stores, a relatively small building adjacent to the giant trading houses. Therefore it is necessary to concentrate all forces.

Concentration of forces
Because being surrounded by large buildings and vast parking lots, the building needs to grow beyond its actual size. The design of the newly adapted shop creates a highly visible and likeable sign with a high recognition factor which serves as a unique business card for E-Tech. The building clearly stands out from the environment because of its shape and colour.

Unmistakable silhouette
The silhouette of the attached north-light roofs and overhanging first floor evokes associations with stacked wooden blocks or exotic characters. This abstract appearance is strongly supported by the use of the monochrome orange colour from the corporate design of the retail chain. So the building itself becomes a logo, a Landmark in an otherwise inter-

East elevation 东立面图

changeable faceless shopping and commercial district.

Actual benefits

The expressive shape brings many advantages for the utilisation: The interior benefits from a clear spatial organization – sales and storage area on the ground floor, offices and meeting rooms are located in the overhanging part above the entrance. The optimal natural lighting of the sales area provided by the north-light roofs not only helps to save energy but also offers a bright, friendly atmosphere. This new airy feeling of space is also characterised by the ceiling height of 6 metres in some parts of the sales floor and a number of new potential views.

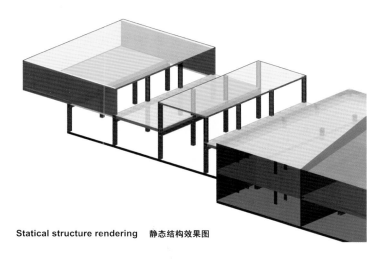

Statical structure rendering 静态结构效果图

1. Northeast exterior view
2. Shopfront design

1. 建筑东北侧外观
2. 店面设计

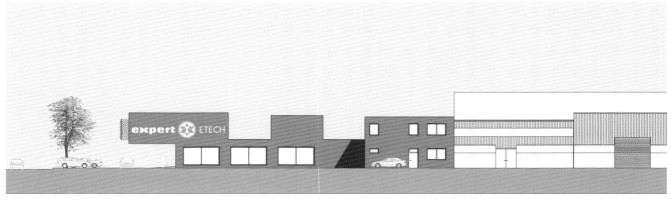

North elevation 北立面图

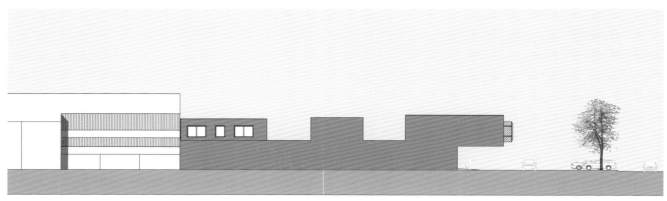

South elevation 南立面图

First floor plan   二层平面图

1. Entrance — 1. 入口
2. Cash Desk — 2. 收银台
3. Small Appliances — 3. 小家电
4. Home Entertainment — 4. 家庭娱乐设施
5. Kitchen Appliances — 5. 厨房电器
6. Laundry & Clothing Care — 6. 洗衣和衣物护理
7. Restrooms — 7. 洗手间
8. Costumer Service — 8. 客户服务
9. Repair Shop — 9. 维修处
10. Storeroom — 10. 仓库

Ground floor plan   一层平面图

11. Store Manager — 11. 经理办公室
12. Staff Room — 12. 员工办公室
13. Storeroom — 13. 仓库

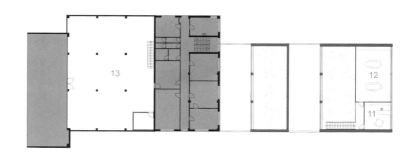

1. Interior rendering
2. North exterior view in night

1. 室内效果图
2. 建筑北侧外观夜景

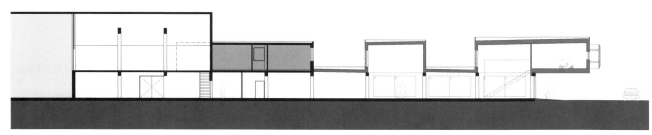

Section 剖面图

Quick realisation

The realisation of the redesign was also extremely cost conscious and was implemented in just 7 weeks of construction. The roof of the old hall was removed and replaced by new prefabricated wall and roof elements, which were mounted on the existing prefabricated concrete hall. The new insulated metal facade encloses the existing as well as the new parts of the building and combines them into a homogeneous whole, and brings the energy standard of the building up to date.

在店铺和购物中心林立，竞争激烈的商业区，吸引路人注意的一个关键挑战是如何给潜在顾客留下深刻印象。当大卫对巨人时，这一点显得更为重要：当备货充足的当地零售连锁店对阵全球连锁店，或者一个相对较小的建筑毗邻巨大贸易公司时，有必要集中所有力量。

集中力量

因为被大型建筑物和广阔的停车场包围，建筑的需要超出其本来大小。新改建的店铺设计带有一个非常明显的可爱商标，作为 E-Tech 公司独特的名片，具有很高的识别度。店铺从周围环境中脱颖而出正是因为它的形状和颜色。

明确无误的轮廓

堆积的木块或异国情调的字符将附带的北方采光屋顶和悬伸的一楼的轮廓关联了起来。公司零售连锁店的这个抽象的外观设计只使用了单色——橘色，因此，这个店铺本身就变成了一个商标，在这个面目模糊的商业区成了一个"地标"。

实际的利益

有表现力的建筑外形能够带来很多实用性：室内有一个清晰的空间布局——底层是销售和储藏区，办公室和会议室位于一层突出的部分。销售区域最佳自然采光由北向采光的屋顶提供，不仅节约能源，还营造了一个明亮而温馨的气氛。这个清新的空间最具特点的是部分销售区的天花板高度达到 6 米，还有大量有潜质的亮点。

快速实现

重新设计的实现也非常耗费精力，在短短 7 周内完成了施工。老厅的屋顶被拆除，取而代之的是新的预制墙体和屋顶元素，它们被安装在现有的预制混凝土大厅。新的绝缘金属幕墙包围了现有的以及新的楼体部分，并把它们整合成一个同质的整体，使建筑达到最新的能效标准。

Ferrari Factory Store
法拉利工厂店

Key words: A Perfect Demonstration of Brand Colour

In many fields a colour has an exclusive meaning. In the world of automobiles, red is definitely for Ferrari. For the first time in the history of Ferrari Stores an entire building has been designed to accommodate the store. The designer encloses the red body in a curved face without mounting posts and bars. Passers-by will immediately notice this distinctive space and recognise the brand.

主题词：品牌代表色的极致彰显

许多领域有些颜色只有一种象征意义，如果要在汽车世界中寻找一种颜色来印证一个品牌，那么非红色的法拉利莫属，该项目是法拉利有史以来第一次以单体建筑的形式来实现其销售门店，设计师将红色的主体建筑囊括在一个弧形玻璃面板组合而成的无框锚固系统之中，路过该街区的人们都会注意到这一特色空间并在最初就能识别该品牌。

Completion date: 2009
Location: Serravalle Scrivia, Italy
Designer: Iosa Ghini Associati
Photographer: Courtesy of Gianluca Grassano
Construction area: 370sqm

Design Points:

Ferrari Factory Store of Serravalle Scrivia, entirely designed by Iosa Ghini Associates, is located outside McArthur Glen Outlet in Serravalle Scrivia. For the first time in the history of Ferrari Stores an entire building has been designed to accommodate the store. The building enjoys a privileged position as one of the first structures of the Outlet visible from the main parking area and access roads, for this reason it was designed with an exterior that immediately identifies it at "Ferrari space".

The building of approximately 370m² is characterised by a large glass gallery that recalls the image and feel of Formula One box, immediately projecting visitors in the Ferrari world.

From a technical point of view the glass gallery is highly innovative with a curved face without mounting posts and bars that permit total visibility inward and outward. The curved glass panels are assembled using a frameless anchor system, i.e. without mounting supports but with clips that guarantees the perception

of material continuity between the plate glasses and provides lightness to the whole system.

Climate control of the glass gallery is effected by a system of air circulation that takes advantage of the motion of air convection, allowing for the passive cooling by natural induction. This natural system is supported by a forced air system that may be activated when climate conditions require it. In addition, the exterior glass envelope is treated with special UV protection films as well as a screen prints that reduces sun rays for energy savings as required by national standards for the sector.

Beyond the glass gallery there is the commercial space. As in all Ferrari Stores the merchandise areas are: the zone for Ferrari fans is designed with aluminium slats of high flexibility; in the luxury zone the display windows use soft materials, brushed leather and polished lacquer; in the children's zone both systems are integrated: slats

1. Entrance
2. Northwest exterior view
3. The car in the display area

1. 入口
2. 建筑西北侧外观
3. 展示区的赛车

1. Northwest exterior view
2. Interior view and cashier

1. 建筑西北侧外观
2. 室内设计和收银台

East elevation　东立面图

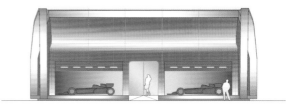

West elevation　西立面图

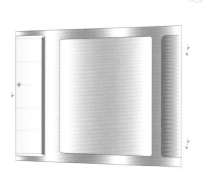

Aerial floor plan　鸟瞰平面图

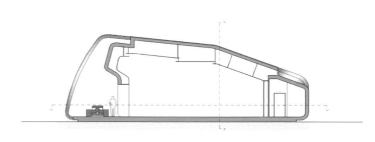

Section　剖面图

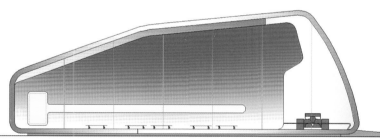

North elevation 北立面图

位于塞拉瓦莱斯克里维亚的法拉利工厂店，完全由尤萨·基尼事务所设计，位于麦克亚瑟格伦直销店外面。这是法拉利专卖店历史上第一次被建成一座独栋大楼。建筑所处的位置十分有利，从停车场和直销店入口处就能看到。因此，建筑的外部设计让人一眼就能看出里面是"法拉利空间"。

建筑总面积约370平方米，巨大的玻璃画廊令人想起了F1方程式赛车，立即把客人带到法拉利的世界。

从技术角度来看，玻璃展廊是一种高度创新。它采用了弧形表面，没有固定柱和隔断，保证了展廊内外的完全通透性。弧形玻璃面板使用无框锚固系统组装，即没有安装支架，也没有用夹子固定，保证了平板玻璃材料的连续性，并且让整个系统显得明亮干净。

玻璃展廊内的气候控制受一个空气循环系统影响。该系统利用空气对流运动实现了被动制冷。这个自然系统由加压气流系统所制成，在必要时可以被激活。此外，外墙玻璃表面以特殊的紫外线防护膜进行了处理，并且配有丝网印花，较少了太阳辐射，实现了行业国家节能标准。

除玻璃展廊外，工厂店还设有商业空间。与所有法拉利店铺的售货区相同：为法拉利粉丝设计的售货区采用铝制板条进行设计，具有高度灵活性；奢华售货区的展示窗采用了柔软的材料、精致皮革和亮漆；儿童区结合了两种设计，铝板和展示窗都采用了黄色亮漆。造型假吊顶勾画出整个售货区的轮廓，引导顾客前行。这些区域的设计都采用了各个法拉利旗舰店标志性设计，彰显了法拉利的品牌特征。

图形风格是项目必不可少的一部分。通过尤萨·基尼的设计，项目实现了三个维度的结合，能够捕获人的所有感官，通过实物和图像传递了令人着迷的概念。

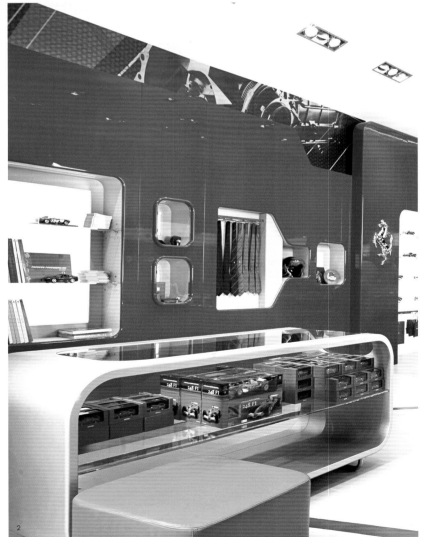

1. Interior view and cashier
2,3. Interior showcase

1. 室内设计和收银台
2、3. 室内展示区

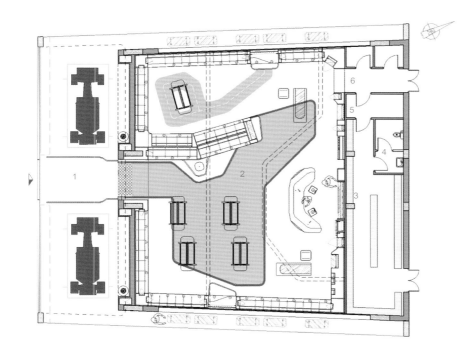

Ground floor plan
1. Glass gallery
2. Store
3. Storage
4. WC
5. Restroom
6. Dressing room

一层平面图
1. 玻璃展览廊
2. 商店
3. 仓库
4. 洗手间
5. 休息室
6. 更衣室

CHAPTER 3 **Cases Study** | Detached Commercial Architecture

Completion date: 2010
Location: Maharashtra, India
Architects: Architect Zarir Mullan and Seema Puri-Mullan
Photographer: Vinesh Gandhi
Construction area: 975sqm

Jewellery Showroom for Batukbhai & Sons

Batukbhai & Sons 珠宝店

Key words: A Well-arranged Structure and Three-dimensional Colourful Lights
The façade is collaged with dark brown reflective glass and dark brown, champagne and cream aluminium panels. These materials with different textures and colours form horizontal and vertical grids according to the building's shape. Lighting design also chooses different sources: the ground floor is lightened with warm golden lighting, while the upper floors' lighting colours are changing all the time, which add fluidity to the whole building. Globe light fittings in silver and copper colour create a space-like atmosphere.

主题词：结构层次分明、灯光色彩立体流动
这个项目的外立面采用深褐色的反光玻璃，深褐色、香槟色和乳白色的铝板，这些不同质感和颜色的材料沿着建筑形状不规则分布形成了水平和垂直的网格。灯光设计也采用了不同的光源，建筑的一层空间的照明是金黄色的暖光源，二层以上的空间光色会发生变化，在夜晚时分会让整个建筑产生一种流动性；内部空间的照明采用了大量的银色和铜色闪耀的球形灯，仰望时有在太空的感觉。

Design Points:

Right from time immemorial as long back as the Indus valley civilisation, Indian women have worn jewellery with pride to set them apart from their counterparts across the globe. This tradition then gets handed down the generations so that it never leaves the family.

Thus Jewellery in India is serious business and a sound investment. The designers had to create a showroom for the oldest Jewellers in the city of Nagpur in central Maha-rashtra India. So as not to lose tradition but rather infuse it with a touch of the modern to use the amazing location in the all important street square to its full potential, to re-energise and almost revitalise the entire square after decades of slumber.

We carved out a 30 feet high atrium space which helped to unify all the levels, create a dramatic triple height entry and most importantly it helped to bring

sunlight into the showroom.

There were three major divisions – gold, diamonds and silver, since the gold section is the largest, and the most popular, it automatically had to be at the ground entrance level, also this was the largest floor plate. The first floor had the diamond section while the silver section was at the second floor. Each floor also had a Director's cabin and a small jewellery lounge for the exclusive customers. Vertical connectivity is provided by the scenic elevator situated at the far end opposite the entrance. The third & fourth floor are the administration sections.

Front exterior elevation 正立面图

1. Night view with the façade blue lighting
2. Exterior view
1. 建筑夜景，配有蓝色灯带
2. 建筑外观

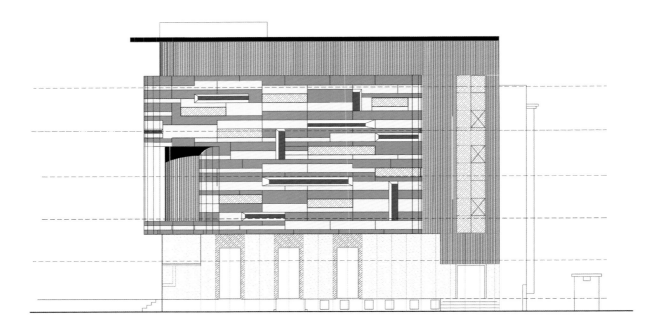

Exterior side elevation
侧立面图

Unique approach
The colour palate was warm with a lot of deep brown wood, travertino marble clad columns with edges of polished copper and a lot of bronze mirror. The desire being to create understated elegance, yet let the customer be transported to another world. To create this ethereal feeling, the designers actually suspended large shiny glass, globe light fitting in the atrium space in silver and copper colour, which when you see from below appear to be floating in space.

The lighting is such that in the evening, the finned masses in the first layer have a deep yellow glow while the niches continuously change colour, this adds fluidity and movement to the built form. The various layers are like the manifestation of the human mind, the deeper ones like the bronze glass signify his deep yet incoherent thought and how slowly these change into more coherent ones, and finally when he emerges with a clear vision of what is his goal, is actually represented, by the luminescent glowing horizontal and vertical slits.

Responsibility
The designers have optimised the daylight into the interiors by carving out a 3-level atrium space.

The intense heat in Nagpur is balanced & kept out by increasing the wall area and using an insulation space between the main building wall and the exterior cladding. The exterior lighting is also done fully in LEDs so as to be completely energy efficient.

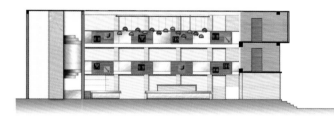

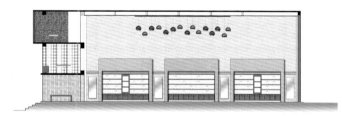

Longitudinal section through triple height showroom
三层楼高的展览区纵剖面

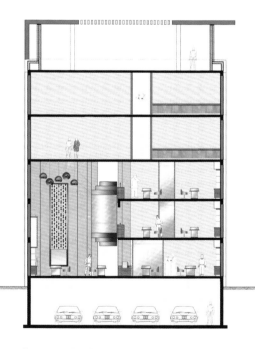

Cross section through showroom
展览区横剖面

1. Night view with the façade red lighting
2. Display with seating

1. 建筑夜景，配有红色灯带
2. 珠宝展示区设有座椅

The other major responsibility in this endeavor was to let the customer be transported to another world. The designers wanted to create an iconic landmark so as soon as you approach the square, the building almost lures you within.

The last responsibility was to infuse the entire square with a positive energy. Any good development is that which energizes the streetscape and context with the surrounding urban environment.

时间倒回印度河流域文明远古时代，印度妇女都骄傲地配戴首饰，他们将自己与其他全球范围内的同类人划分开来。这个传统被代代流传下来的，从来没有离开过。

珠宝行业在印度商业领域有举足轻重的地位，投资规模也很大。我们必须在印度马哈拉施特拉邦中部的城市那格浦尔为那些老珠宝商创造一个珠宝陈列室。为了既不失去传统又能注入一些现代感，我们决定利用街道广场重要的位置，以充分发挥其潜力。项目为沉睡了几十年的广场重新注入了活力。

项目通过设计统一了所有楼层，创造了一个引人注目的3倍高的入口，最重要的是阳光可以照进陈列室。

店铺有三个主要分区，黄金区、宝石区和银饰区。由于黄金区是最大的、最流行的，它自然在底层入口处，这

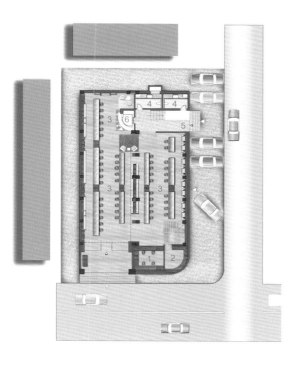

Ground floor plan
1. Cashier
2. Strong room
3. Gold section
4. Toilet
5. Back entrance
6. Life

First floor
1. Lounge
2. Strong room
3. Directors Cabin
4. Diamond section
5. Toilet
6. Back entrance
7. Lift

Second Floor
1. Lounge
2. Strong room
3. Directors Cabin
4. Silver section
5. Toilet
6. Back staircase
7. Lift
8. Store

Third floor
1. Exhibition hall
2. Staff lunch room
3. Box room
4. Accounts
5. Server room
6. Change room/lockers
7. Back staircase
8. Staircase

一层平面图
1. 收银台
2. 保险库
3. 黄金区
4. 洗手间
5. 后门
6. 生活区

二层平面图
1. 休息室
2. 保险库
3. 主管办公室
4. 宝石区
5. 洗手间
6. 后门
7. 电梯

三层平面图
1. 休息室
2. 保险库
3. 主管办公室
4. 银饰品
5. 洗手间
6. 后楼梯
7. 电梯
8. 仓库

四层平面图
1. 展览厅
2. 员工午餐室
3. 储藏室
4. 财务室
5. 服务器机房
6. 更衣室/储物箱
7. 后楼梯
8. 楼梯

1. Top view of display area
2. Display area

1. 俯瞰展示区
2. 展示区

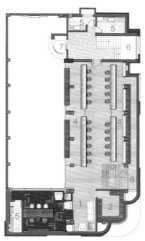
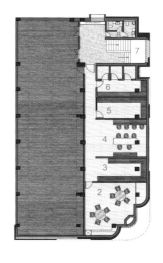

也是最大的一层。宝石区和银饰区分别在二层和三层。每层楼都有一个主管办公室和小型珠宝间供VIP客户休息。垂直连接是位于远端入口对面的观光电梯。第四和第五层是行政主管部门的办公区。

优势
室内是暖色系，使用了深褐色的木材，边缘有抛光铜的洞石大理石包柱和大量青铜镜。我们想要创造低调优雅的环境，让顾客感受到这里的特别。为了营造空灵的感觉，我们在室内悬浮了大型有光泽的玻璃圆球状灯饰，与中庭空间中银和铜的颜色相呼应，所以当你从下面向上看时，会感觉自己仿佛飘浮在太空中。

到了晚上，一层透出很深的黄色光晕，随着壁龛不断改变颜色，增加了建筑的流动性和运动感。不同楼层灯光的变换就像人类心灵的表现，更深层的青铜玻璃表现着深刻但不连贯的感觉，并且展示着如何慢慢变为连贯，最后表达了一个清晰的目标。这些都是通过那些发光的水平和垂直的缝隙来表现的。

责任
设计师通过在内部创造出三层的中庭空间来优化自然光。热量通过增加外墙表皮和使用的来实现平衡和保持，增加的绝缘空间主要是建筑墙体和外部包层之间的空间。外部照明也全部使用LED，以使其达到完全高效节能。

设计师另一个主要责任是让顾客感到被带到另一个世界。他们希望创建一个标志性的建筑，只要你接近广场，这座大楼就会诱惑你进去。

设计师的最后一个责任是将积极的能量注入整个广场。任何能为街景与周围的城市环境带来活力和能量的发展都包含其中。

Completion date: 2009
Location: Maryland, USA
Designer: RTKL Associates Inc.
Photographer: Mark A. Steele
Site area: 344sqm

Peeps & Company™

小鸡糖果旗舰店

Key words: Colourful Design Catering to Children's Interests

The yellow exterior is in line with the owner's brand, which was built on yellow Peeps® marshmallows. Because of the store's position on a sunken plaza, the project is designed in an arc and features transparent, full-height glazing in display windows and the façade. The interior is divided according to the colours of Just Born's various brands, which are visible even from the outside. The coloured tables and chairs outside the store also correspond to the interior layout.

主题词：多彩设计迎合童趣

黄色的外观设计与小鸡糖果品牌的黄色棉花糖相互呼应。加之项目所处的下沉广场的地理位置，项目整体造型呈圆弧形，配以通透的落地玻璃作为橱窗及外墙。建筑内部以其下属品牌的色彩基调分区，在室外即能感受到其中五彩斑斓的世界。店外还设有座位色彩各异的休息桌椅，与室内布局交相呼应。

Design Points:

Seeking to elevate brand awareness, Just Born Inc. selected RTKL to create a retail flagship for its popular line of candy products including Peeps®, Mike And Ike®, Hot Tamales®, Teenee Beanee® and Peanut Chews® candies.

From collaborating on the naming of the store to developing the store's brand positioning, RTKL created a fully customised, branded retail environment to tell the Peeps story. The flagship store Peeps and Company™ opened in National Harbour, Maryland in 2009.

The primary challenge was to communicate the individuality of each product while positioning them all under the Peeps and Company™ identity. To meet this challenge, the design team created a storybook concept set within an experiential environment where the products are stars on a stage. Within this concept, lighting, sound, video technology, interior finishes and materials were carefully selected to bring the candy products to life in an integrated sensory retail experience.

The story begins the moment customers walk through the front doors as they are greeted with a giant glowing Peep physically centred in a modern white backdrop punctuated by individual brand colours divided into four zones. Flanked by a Mike And Ike® bulk candy wall housed in a giant graphic equalizer on one side and a Hot Tamales® "How Hot are you" thermometer on the other, the giant Peep sets the mood with the colour of its glow. Since the brands are driven by the seasons, the theme of the retail environment transitions with the help of thousands of colour-changing LED lights as the seasons progress.

The store is also built to LEED-CI Silver standards. The combination of innovative branding, environmental graphics, interior design features and integrative technology work to create a successful and compelling retail experience.

1. Exterior view at night
2. Interior view
3. The Registers

1. 建筑外观夜景
2. 店铺内部
3. 登记台

1. Hot Tamales' area
2. The showcase
3. Hot Tamales® and Mike and Ike®

1. Hot Tamales'区
2. 产品展示区
3. Hot Tamales® and Mike and Ike®区

为了提升品牌知名度，Just Born 公司委托 RTKL 设计公司为它旗下的糖果产品品牌 Peeps、Mike And Ike、Hot Tamales、Teenee Beanee 和 Peanut Chews 打造一家旗舰店。

从旗舰店的名称到品牌定位，RTKL 设计公司量身打造了一个完整的品牌销售环境，讲述了小鸡糖果的故事。2009 年，小鸡糖果旗舰店在马里兰州的国家港口正式开业。

项目的主要挑战是如何在保持旗舰品牌统一性的前提下展现子品牌产品的特点。为了实现这一目标，设计团队以故事书的形式打造了一个体验环境，而产品就宛如舞台上的明星。以此为理念，项目的灯光、音效、视频技术、室内装饰和材料的选择都力求赋予糖果产品鲜活的生命力，使其在体验环境中大放异彩。

故事从消费者一走进正门就已经展开：一只巨大的发光小鸡在白色布景正中欢迎消费者。旗舰店内部以子品牌的色彩划分为四个区域。店内两侧的墙面上分别是 Mike And Ike 以图形均衡器为背景的散装糖果墙和 Hot Tamales 标有"How Hot are you"（到底有多火热）的巨型温度计。中央巨大的发光小鸡奠定了空间的色彩基调。由于各个品牌的产品经常换季，多变的 LED 彩灯可以通过变换来改变室内环境，使其与品牌同步。

旗舰店的设计以 LEED-CI 银奖为标准。创新品牌、环境图形设计、室内设计特征和综合技术相结合，共同营造出富有魅力的零售体验。

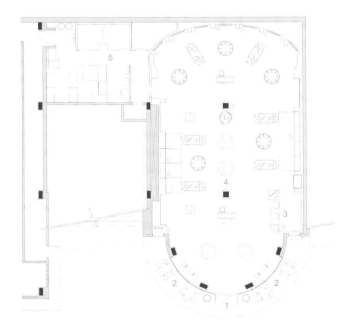

Ground floor plan   一层平面图

1. Entrance   1. 入口
2. Outdoor seating   2. 露天座椅
3. Cashier   3. 收银台
4. Marketplace   4. 卖场
5. Back of house   5. 工作区

CHAPTER 3 **Cases Study** | Detached Commercial Architecture

The development of low-carbon economy and the construction of low-carbon city have become a focus of global urban construction. Nowadays, construction industry is gradually tend for low carbon and energy saving. Various aspects including building materials, equipment manufacture, construction and building operation life cycle can achieve low carbon emission and high efficiency. Low carbon has become an architectural concept acknowledged and promoted by almost all the architects.

发展低碳经济、构建低碳城市已日渐成为全球城市建设的工作重点，如今，建筑行业也逐步实现低碳节能。建筑材料、设备制造、施工建造和建筑物使用的整个生命周期等方面都能实现碳排放减少、提高效能，成为每个建筑师都认同并推崇的建筑理念。

ARCHITECTURE AND SUSTAINABILITY

建筑与可持续发展

4.1 Building Envelope
The building envelope is a key element of an energy-efficient design.

4.1 建筑外壳
建筑外壳是节能设计的主要元素之一。

1. The façade of Municipal Market
2. Bird's eye view of Climate Protection Supermarket

1. 市政市场外立面
2. 气候保护超市鸟瞰图

4.1.1 Roofs

Cool roofs are recommended for roofs with insulation entirely above deck and for metal building roofs. In metal roof building construction, purlins are typically Z-shaped cold-formed steel members, although steel bar joists are sometimes used for longer spans. The thermal performance of metal building roofs with fiberglass blankets is improved by addressing the thermal bridging associated with compression at the purlins. The two types of metal building roofs are standing seam roofs and through-fastened roofs.

Attics and other roofs include roofs with insulation entirely below (inside of) the roof structure (i.e., attics and cathedral ceilings) and roofs with insulation both above and below the roof structure. Ventilated attic spaces need to have the insulation installed at the ceiling line. Unventilated attic spaces may have the insulation installed at the roof line. When suspended ceilings with removable ceiling tiles are used, the insulation needs to be installed at the roof line.

Single rafter roofs have the roof above and ceiling below both attached to the same wood rafter, and the cavity insulation is located between the wood rafters. Continuous insulation, when recommended, is installed to the bottom of the rafters and above the ceiling material.

4.1.2 Walls

Mass walls are defined as those with a heat capacity exceeding 7 Btu/ft^2 · °F. Insulation may be placed on either the inside or the outside of the masonry wall. The greatest advantages of mass can be obtained when insulation is placed on the exterior of the mass. In this case, the mass absorbs internal heat gains that are later released in the evenings when the buildings are not occupied.

Cold-formed steel framing members are thermal bridges to the cavity insulation. Adding exterior foam sheathing as continuous insulation is the preferred method to upgrade the wall thermal performance be-

4.1.1 屋顶

建议顶部有整体绝缘设施的屋顶以及金属建筑屋顶采用冷屋顶。在金属屋顶建筑的建设中，虽然较长的跨度有时会使用钢筋桁架，其檩条通常为Z形冷弯型钢。通过与檩条压迫相关的热桥效应，可以提升带有玻璃纤维层的金属建筑屋顶的热性能。金属建筑屋顶分为两种：立接缝屋顶和全固定屋顶。

阁楼及其他屋顶包括绝缘层整体设在屋顶结构下方（或内部）的屋顶（如阁楼和教堂天花板）和绝缘屋顶结构上下方都设有绝缘层的屋顶。通风的阁楼空间需要在天花板走线上安装绝缘。不通风的阁楼空间可以在屋檐线上安装绝缘。如果采用可移除式吊顶板的吊顶，绝缘需要安装在屋檐线上。

单椽屋顶的屋顶在上方，天花板在下方，二者都附着于同样的木椽，空腔绝缘设在木椽之间。连续绝缘可以安装在椽的底部和天花板材料的上方。

4.1.2 墙壁

蓄热墙的热容量需要在7百万英热/平方尺·华氏以上。可以在砌筑墙的内部或外部设置绝缘。当绝缘设在墙体外层时，可以实现墙体效率的最大化。在这种情况下，墙体所吸收的内部热量将在建筑无人使用的夜晚排出。

冷弯型钢框架构件是空腔绝缘的热桥。建议添加外部泡沫板作为连续绝缘，以此提升墙体的热性能。因为泡沫

cause it will increase the overall wall thermal performance and tends to minimise the impact of the thermal bridging.

Cavity insulation is used within the wood-framed wall, while rigid continuous insulation is placed on the exterior side of the framing. Care must be taken to have a vapour barrier on the warm side of the wall and to utilise a vapour-barrier-faced batt insulation product to avoid insulation sagging away from the vapor barrier.

4.1.3 Floors
Insulation should be continuous and either integral to or above the slab. It should be purchased by the conductive R-value. This can be achieved by placing high-density extruded polystyrene above the slab with either plywood or a thin layer of concrete on top. Placing insulation below the deck is not recommended due to losses through any concrete support columns or through the slab perimeter.

Insulation should be installed parallel to the framing members and in intimate contact with the flooring system supported by the framing member in order to avoid the potential thermal short circuiting associated with open or exposed air spaces.

4.1.4 Slabs
Rigid continuous insulation should be used around the perimeter of the slab and should reach the depth listed in the recommendation or to the bottom of the footing.

When slabs are heated, rigid continuous insulation should be used around the perimeter of the slab and should reach to the depth listed in the recommendation or to the bottom of the footing, whichever is less. Note that it is important to use the conductive R-value for the insulation as radiative heat transfer is small.

板可以增加整个墙体的热性能并缩小热桥效应的影响。

木框架墙内通常使用空腔绝缘，连续刚性绝缘被设置在框架的外侧。必须注意在墙壁的暖面添加防潮层并利用带有防潮面的条毯式绝缘产品来避免绝缘层松垂脱离防潮层。

4.1.3 地面
绝缘层应当是连续的，可以与楼板成为一体，也可以设在楼板之上。应根据传导性热阻购买绝缘材料。可以在楼板上使用高密度挤塑聚苯乙烯板，上方铺设胶合板或薄水泥层。不建议在楼板下方设置绝缘，因为混凝土支柱或楼板边缘都会造成热损失。

绝缘层应当与框架构件平行，并与框架构件所支撑的地面系统紧密相连，以避免产生与开放或暴露的气室相关的热短路。

4.1.4 楼板
楼板边缘应当采用连续刚性绝缘并达到推荐的深度或到达楼板底脚的底部。

如果楼板被加热，楼板边缘所采用严密的连续绝缘应达到推荐的深度或到达楼板底脚的底部，取二者的最小值。注意：由于辐射传热量较小，一定要使用传导性热阻来进行绝缘。

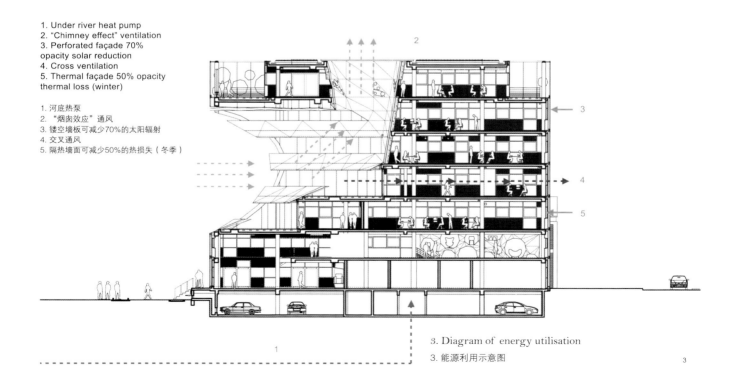

1. Under river heat pump
2. "Chimney effect" ventilation
3. Perforated façade 70% opacity solar reduction
4. Cross ventilation
5. Thermal façade 50% opacity thermal loss (winter)

1. 河底热泵
2. "烟囱效应"通风
3. 镂空墙板可减少70%的太阳辐射
4. 交叉通风
5. 隔热墙面可减少50%的热损失（冬季）

3. Diagram of energy utilisation
3. 能源利用示意图

4.1.5 Doors

A U-factor of 0.37 corresponds to an insulated double-panel metal door. A U-factor of 0.61 corresponds to a double-panel metal door. If at all possible, single swinging doors should be used. Double swinging doors are difficult to seal at the centre of the doors unless there is a centre post. Double swinging doors without a centre post should be minimised and limited to areas where width is important. Vestibules can be added to further improve the energy efficiency.

Roll-up or sliding doors are recommended to have R-4.75 rigid insulation or meet the recommended U-factor. When meeting the recommended U-factor, the thermal bridging at the door and section edges is to be included in the analysis. Roll-up doors that have solar exposure should be painted with a reflective paint (or high emmissivity) and/or should be shaded. Metal doors are a problem in that they typically have poor emmissivity and collect heat, which is transmitted through even the best insulated door, causing cooling loads and thermal comfort issues in the space. If at all possible, use insulated panel doors over roll-up doors, as the insulation values can approach R-10 and provide a tighter seal to minimise infiltration.

4.1.6 Vertical Glazing

Window-wall ratio is the percentage resulting from dividing the total fenestration, including glazed doors, by the gross exterior wall area. For any area less than 40%, the recommended values for U-factor and solar heat gain coefficient ratio contribute to the 30% savings target of the entire building. A reduction in the overall fenestration area will also save energy, especially if glazing is significantly reduced on the east and west façades.

4.1.5 门

一扇双层金属板绝缘门的传热系数为 0.37。一扇双层金属板门的传热系数为 0.61。双开门很难实现密封，除非在两扇门板中间安装一个立柱。如果没有特殊的宽度要求，应尽量避免使用没有中央立柱的双开门。可以增加门厅，以进一步提高能源效率。

建议卷帘门或拉门的刚性绝缘达到 R-4.75 或符合推荐的传热系数。应当推荐的传热系数中对门与墙面边缘的热桥效应进行分析。能受到日光照射的卷帘门的辐射系数通常较低，热量可以贯穿绝缘良好的门板，造成室内制冷负荷及热舒适度问题。如有可能，应在卷帘门外安装绝缘门板，使其绝缘值接近 R-10，实现更严密的密封，实现热渗透的最小化。

4.1.6 垂直玻璃装配

窗墙比即窗户开口的总面积（包含玻璃门）除以总外墙面积的百分比。当窗墙比小于 40% 时，传热系数和太阳能得热比例的推荐值可以实现整座建筑节能目标的 30%。减少整体开窗面积也能实现节能，特别是减少东、西两面墙上的玻璃装配。

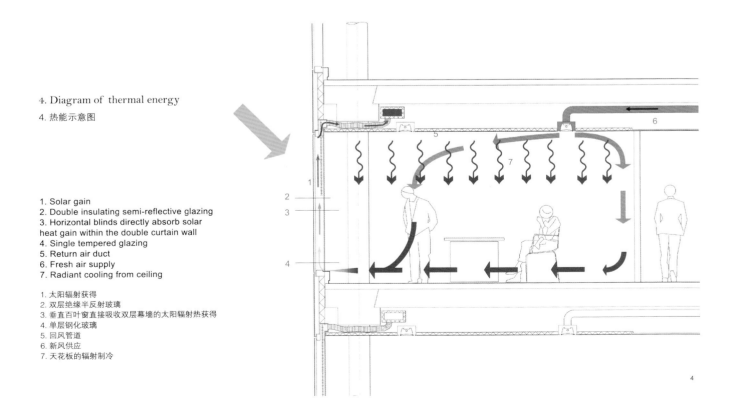

4. Diagram of thermal energy
4. 热能示意图

1. Solar gain
2. Double insulating semi-reflective glazing
3. Horizontal blinds directly absorb solar heat gain within the double curtain wall
4. Single tempered glazing
5. Return air duct
6. Fresh air supply
7. Radiant cooling from ceiling

1. 太阳辐射获得
2. 双层绝缘半反射玻璃
3. 垂直百叶窗直接吸收双层幕墙的太阳辐射热获得
4. 单层钢化玻璃
5. 回风管道
6. 新风供应
7. 天花板的辐射制冷

4.1.7 Skylights

Skylights provide increased daylight and a potential reduction in lighting energy consumption at the expense of increasing cooling loads in warmer climates and increasing heating loads in cooler climates. To achieve the lighting energy savings, the lighting in fixtures within ten feet of the skylight edge must have automatic controls that dim the lighting in response to available daylight.

Hot Climates

• Use north-facing clerestories for skylighting whenever possible in hot climates to eliminate excessive solar heat gain and glare. Typically, north-facing clerestories have one-sixth the heat gain of skylights.
• Reduce thermal gain during the cooling season by using skylights with a low overall thermal transmittance (U-factor). Insulate the skylight curb above the roofline with rigid c.i.
Shade skylights with exterior/interior sun control such as screens, baffles, or fins.
Use smaller aperture skylights in a grid pattern to gain maximum usable daylight with the least thermal heat transfer.

Moderate and Cooler Climates

• Use either north- or south-facing clerestories for skylighting but not east- or west-facing ones. East-west glazing adds excessive summer heat gain and makes it difficult to control direct solar gain. Clerestories with operable glazing may also help provide natural ventilation in temperate seasons when air conditioning is not in use. Typically, north-facing clerestories have one-sixth the heat gain of skylights.

• Reduce summer heat gain as well as winter heat loss by using skylights with a low overall thermal transmittance. Use a skylight frame that has a thermal break to prevent excessive heat loss/gain and winter moisture condensation on the frame. Insulate the skylight curb above the roofline with rigid c.i.
Shade south-facing clerestories and skylights with exterior/interior sun control such as screens, baffles, or fins.
Use skylights with smaller apertures in a grid pattern to gain maximum usable daylight with the least thermal heat transfer. Do not exceed maximum prescribed glazing area.
Splay skylight opening at 45° to maximise daylight distribution and minimise glare.

4.1.8 Obstructions and Planting

Adjacent taller buildings and trees, shrubs, or other plantings are effective for shading glass on south, east, and west facades. For south-facing windows, remember that the sun is higher in the sky during the summer, so shading plants should be located high above the windows to effectively shade the glass. The glazing of fully shaded windows can be selected with higher SHGC ratings without increasing energy consumption. The solar reflections from adjacent buildings with reflective surfaces (metal, windows, or especially reflective curtain walls) should be considered in the design. Such reflections may modify shading strategies, especially on the north façade.

4.1.9 Passive Solar

Passive solar energy-saving strategies should be limited to non-sales and non-office spaces, such as lobbies and circulation areas, unless these strategies are designed so that workers and customers do not directly view interior sun patches or see them reflected on merchandise or work surfaces. Consider heat-absorbing blinds in cold climates or reflective blinds in warm climates. In spaces where glare is

4.1.7 天窗

天窗能提供额外的日光，减少照明能源消耗。但是，天窗在气候温暖时会增加制冷负荷，在气候寒冷时则会增加采暖负荷。为了实现照明能源的节约，距天窗10英尺（约3.05米）范围内的灯具必须具备自动控制系统，可根据可利用日光量进行调光。

炎热气候

• 尽量在炎热气候使用朝北的侧天窗，以避免过度的太阳辐射得热增量和眩光。通常情况下，朝北的侧天窗的热增量是天窗的六分之一
• 通过使用低传热系数的天窗来减少制冷季节的热增量。利用以下方式来为屋顶线上方的天窗侧沟进行绝缘：
采用室内/室外阳光控制装置为天窗遮阳，例如遮阳板、挡板或散热翅片。
将较小开口的天窗呈网格状排列，以最少热传递量获得最多的可用日光。

温和及凉爽的气候

• 使用南北朝向的侧天窗，而不是东西朝向的。东西朝向的玻璃装配会增加过度的夏季热增量并增大控制直接太阳辐射热增量的难度。在温和的气候，如果不使用空调，可控制的侧天窗可以帮助提供自然通风。通常情况下，朝北的侧天窗的热增量是天窗的六分之一
• 利用低传热系数的天窗来减少夏季热增量和冬季热损失。利用带有断热装置的天窗窗框来防止窗框上的过度热增量/损失。利用以下方式来为屋顶线上方的天窗侧沟进行绝缘：
采用室内/室外阳光控制装置为天窗遮阳，例如遮阳板、挡板或散热翅片。
将较小开口的天窗呈网格状排列，以最少的热传递量获得最多的可用日光。
将天窗开口倾斜45°，以最大化日光分布、减少眩光。

4.1.8 障碍物和植物

紧邻高楼以及利用树木、灌木或其他植物都能有效地对东、西、南三面的玻璃进行遮阳。对南窗户来讲，需要注意夏季的太阳更高，遮阳植物应置在高于窗口的地方进行有效遮阳。在不增加能源消耗的前提下，进行全面遮阳的窗户可以采用太阳得热系数较高的玻璃。设计应当考虑相邻建筑的反光表面（金属、窗户，特别是反光幕墙）所反射的日光。这些反光可能会影响遮阳策略，特别是北墙的遮阳策略。

4.1.9 被动式太阳能

被动式太阳能节能策略应限制在非销售、非办公的区域，例如大厅、流通区域等。除非这些策略的设计让工作人员或消费者不能直接看到室内阳光斑块或者看不到太阳斑块反射在商品或工作台面上。考虑在寒冷气候

not an issue, the usefulness of the solar heat gain collected by windows can be increased by using massive thermally conductive floor surfaces, such as tile or concrete, in locations where the transmitted sunlight will fall. These floor surfaces absorb the transmitted solar heat gain and release it slowly over time, to provide a more gradual heating of the structure. Consider low-e glazing with exterior overhangs.

4.2 Lighting

In small retail buildings, lighting plays a significant role in the energy consumption of the building; its impact becomes more pronounced in cooling-dominated climates. Lighting will often be designed after construction of the shell. If possible, the lighting loads should be specified before selection of the HVAC systems in order to select the size and system type for the most efficient and cost-effective approach.

4.2.1 Daylighting

Daylight in buildings can save energy if the electric lighting is switched or dimmed in response to changes in daylight levels in the store. Automatic lighting controls increase the probability that daylighting will save energy. It is also important that heat gain and loss through glazing be controlled. In addition, glare and contrast must be controlled so occupants are comfortable and will not override electric lighting controls.

Daylighting utilises light from the sky, not the direct sun. Patches of direct sunlight in the sales area will create unacceptable brightness and excessive contrast between light and dark areas.

• Use exterior and interior sun control devices. Exterior sun control and overhangs help reduce both direct sun penetration and heat gain from vertical glazing surfaces.
• Use continuous exterior overhangs and interior horizontal blinds or shades on south-facing glazing.
• Use interior vertical slat blinds or shades on east- and west-facing glazing and as required for northeast or northwest façades.
• An exterior overhang needs to be deep enough to shield windows above the light shelf (if used) from direct sun. The light shelf, or the overhang if the light shelf is not used, should also be deep enough to shield windows below the shelf from direct sun.
• For "top-lighting," use north-facing clerestories to avoid direct sun.
• For skylights, use light-reflecting baffles and/or diffusing glazing to control direct sun. Note that diffusing skylights can cause glare when the sun hits them.

使用吸热百叶窗，温暖气候使用反射百叶窗。在没有眩光问题的空间，可以通过在透射阳光照射的地面采用大面积的导热地面材料（如地砖、水泥）来增加窗户收集的太阳热增量。这些地面能吸收透射的太阳热增量，并对其进行缓慢地释放，为建筑结构提供持续的供暖。考虑在室外屋檐使用低辐射玻璃。

4.2 照明
在小型零售业建筑中，照明在建筑的能源消耗中占有主要位置；它的影响在制冷为主的气候中更为重要。照明设计通常在建筑封顶之后进行。应尽量在选择空调系统之前确定照明负荷，以便做出最合适、最节能、最划算的选择。

4.2.1 日光照明
如果随着日光等级将电灯关掉或调暗，建筑的日光照明可以实现节能。自动照明控制提高了日光照明的节能概率。控制穿透玻璃引起的热增量和热损失同样重要。此外，必须控制眩光和反差，使建筑内的人感到舒适，不至于强行改变照明控制。

日光照明使用来自天空的光线，而不是直接的阳光。在销售区域，直接阳光的斑块会造成难以接受的亮度并且在明暗区域之间形成强烈的对比。

• 采用室内外太阳控制设备。室外太阳控制和屋檐能够帮助减少垂直玻璃表面的直接太阳渗透和热增量
• 在朝南的玻璃上使用连续的室外屋檐和室内水平百叶窗或遮阳板
• 室外屋檐需要足够的深度来保护光架上方的窗户不受阳光直射。光架（如果没有光架，则是屋檐）同样应有足够的深度来保护光架下方的窗户不受阳光直射
• 顶部照明应使用朝北的侧天窗来避免阳光直射
• 天窗应采用反光挡板及/或漫射玻璃来控制阳光直射
注意：当太阳照射时，漫射天窗可能会造成眩光。

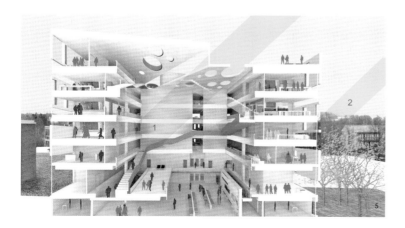

5. Diagram of lighting

5. 光照示意图

1. Naturally daylit atrium
2. Light, filtered through surrounding spaces

1. 自然采光中庭
2. 通过周边空间过滤的光线

4.2.2 Interior Electric Lighting
The daylight zone is the area of the skylight plus the floor-to-ceiling dimension in all directions from the edge of the skylight. The daylight zone at the perimeter is 15 ft deep and 2 feet wider than the window.

Dimming controls. In merchandise sales areas, continuously dim rather than switch electric lights in response to daylight to minimise customer/employee distraction. Specify dimming ballasts that dim down to at least 20% of full output. Automatic multilevel daylight switching may be used in non-sales environments such as hallways, storage areas, restrooms, lounges, etc. To maintain an adaptation zone (high light level during the daylight hours), dimming of the luminaires adjacent to the entry is not recommended. Control luminaires in groups around skylights, and if using a lighting system that provides an indirect component, do not dim below 20% to maintain a brightness balance between skylights and surrounding ceiling. If daylight zones overlap, a single control zone may be used. The daylighting control system and/or photosensor should include a five-minute time delay or other means to avoid cycling caused by rapidly changing sky conditions and a one-minute fade rate to change the light levels by dimming.

4.2.3 Decorative Lighting
Decorative lighting (wall sconces and pendant fixtures) can add visual interest and focus to the space, especially at the sales transaction area. These fixtures are included in the tabulation of the base LPD, and consideration should be given to energy-efficient solutions, including compact fluorescent, ceramic metal halide (CMH), and light-emitting diode (LED) lighting.

4.2.4 Lighting Controls
Factory setting of calibrations should be specified when feasible to avoid field labour. Lighting calibration should be performed after furniture installation but prior to occupancy to ensure user acceptance.

Use an astronomical time switch for all exterior lighting. Astronomical time switches are capable of retaining programming and time settings during loss of power for a period of at least ten hours. If a building energy management system is being used to control and monitor mechanical and electrical energy use, it can also be used to schedule and manage outdoor lighting energy use. Turn off exterior lighting not designated for security purposes when the building is unoccupied.

4.3 HVAC Equipment and Systems
HVAC systems are key elements. How systems are installed affect how efficiently they can be serviced and how well they will perform.

4.3.1 HVAC System Types
The packaged-unit systems and split systems with a refrigerant-based direct expansion system for electric cooling and heating by means of one of the three following options:
Option 1: Indirect gas-fired heater
Indirect gas-fired heaters use a heat exchanger as part of the factory-assembled unit to separate the burner and products of combustion from the circulated air.
Option 2: Electric resistance heater
Electric resistance heaters can be part of the factory-assembled unit or can be installed in the duct distribution system.
Option 3: Heat pump unit

4.2.2 室内电气照明
日光区域指天窗区域及从天窗边缘到各个方向的地面到天花板的维度。日光区域周长纵深 15 英尺（约 4.57 米），比窗户宽 2 英尺（约 0.61 米）。

调光控制。在商品销售区域，连续的调光——而不是开关电灯——能够避免干扰到消费者和工作人员。选择至少能够调整到完整输出的 20% 的调光镇流器。可以在走道、仓储区、洗手间、休息室等非销售区域使用多级自动日光开关。为了维持适应区域（在日照时间保持高亮度），不建议在入口处采用调光照明设备。对天窗四周的照明器进行分组控制。如果使用了具有间接组件的照明系统，不要将光调到 20% 以下，以保证天窗和周围天花板之间的亮度平衡。如果日光区域部分重叠，可以采用一个单一控制区。日光控制系统及 / 或感光器应当包含一个 5 分钟的延时或其他方式来避免快速转换天空环境所带来的冲击，还应包含一个 1 分钟的淡出率来通过调光变化光照等级。

4.2.3 装饰照明
装饰照明（壁灯和吊灯）能够增强视觉趣味和空间的关注度，特别是在销售区域。这些灯具在照明功率密度表格里都有体现，需要注意它们的节能手段，包括紧凑型荧光灯、陶瓷金属卤素灯（CMH）和发光二极管（LED）。

4.2.4 照明控制
如果可行，应当明确校准出厂设置，以避免安装问题。应当在家具安装完毕后、建筑投入使用前进行照明校准，以保证使用者的接受度。

使用天文时间开关对所有室外照明进行控制。在无电源的情况下，天文时间开关可以保留程序设置和时间设定至少 10 小时。如果建筑采用能源管理系统对机械和电气能源使用进行控制和监控，也可以用它来预订和管理室外照明能源使用。当建筑无人使用时，关闭建筑外部的非安全用途照明。

4.3 空调设备和系统
空调系统是设计的关键因素。系统的安装影响着它们的效率和性能。

4.3.1 空调系统的类型
采用以制冷剂为基础的直接膨胀式制冷加热系统的机组和分散式系统有以下三种方案：
方案 1：间接燃气加热器
间接燃气加热器以热交换器作为出厂机组的一部分，使燃烧器和燃烧产物与流通空气隔开。
方案 2：电阻加热器
电阻加热器可以是出厂机组的一部分，也可以安装在管道分配系统。
方案 3：热泵机组
在除霜循环中，热泵机组的辅助热源也可以用来进行空间供暖。既可以电力驱动，也可以燃气驱动。

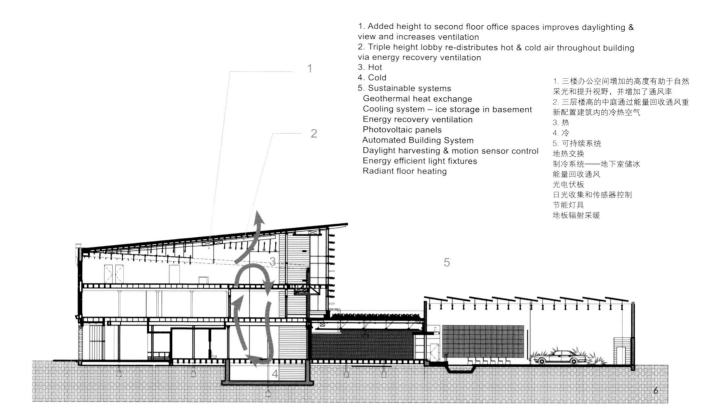

1. Added height to second floor office spaces improves daylighting & view and increases ventilation
2. Triple height lobby re-distributes hot & cold air throughout building via energy recovery ventilation
3. Hot
4. Cold
5. Sustainable systems
   Geothermal heat exchange
   Cooling system – ice storage in basement
   Energy recovery ventilation
   Photovoltaic panels
   Automated Building System
   Daylight harvesting & motion sensor control
   Energy efficient light fixtures
   Radiant floor heating

1. 三楼办公空间增加的高度有助于自然采光和提升视野，并增加了通风率
2. 三层楼高的中庭通过能量回收通风重新配置建筑内的冷热空气
3. 热
4. 冷
5. 可持续系统
   地热交换
   制冷系统——地下室储冰
   能量回收通风
   光电伏板
   日光收集和传感器控制
   节能灯具
   地板辐射采暖

6. Diagram of hot and cold air circulation
6. 冷热空气循环示意图

The auxiliary heat source for heat pump units may also be used to supply heating to the space during the defrost cycle and can be either electric or gas.

Where variable air volume systems are used, the refrigeration system requires reduced capacity in response to reduced load. The package unit should be designed to maintain the required apparatus dew point for humidity control. The controls of a variable air volume system should be arranged to reduce the supply air to the minimum set-point for ventilation before tempering of the air occurs. Variable-speed drives should be considered as an option to reduce airflow and fan/motor energy.

4.3.2 Cooling Equipment Efficiencies and Heating Equipment Efficiencies

Any safety factor should be applied cautiously and only to a building's internal loads to prevent oversizing of equipment. If the unit is oversized and the cooling capacity reduction is limited, short-cycling of compressors could occur and the system may not have the ability to dehumidify the building properly. Include the cooling and heating load of the outdoor air to determine the total cooling and heating requirements of the unit. In determining cooling requirements, the sensible and latent load to cool the outdoor air to room temperature must be added to the building cooling load. For heating, the outdoor air brought into the space must be heated to the room temperature and the heat required added to the building heat loss.

在使用变风量系统时，制冷系统需要通过降低容量来降低负荷。机组的设计应保持必要的露点温度来进行湿度控制。在空气调节开始之前，变风量系统的控制应当能够将送气量降到最小的通风设定值。建议使用变速驱动来减少气流和风机／电机能耗。

4.3.2 制冷设备效率与加热设备效率

必须应对建筑内部负荷谨慎应用安全系数，以防止设备的尺寸超标。如果机组尺寸超标，而制冷能力缩减有限，则可能造成压缩机短路，致使系统无法为建筑进行适当的除湿。在计算机组的整体制冷和加热要求时，将户外空气的制冷和加热纳入考虑。在决定制冷需求时，将户外空气冷却到室温的显热负荷和潜热负荷必须加入建筑制冷负荷之中。在加热方面，需要添加进入室内的户外空气被加热到室温的负荷以及补偿建筑热损失的负荷。

4.3.3 Control Strategies

Control strategies can be designed to help reduce energy. Time-of-day scheduling is useful when it is known which portions of the building will have reduced occupancy. Control of the ventilation air system can be tied into this control strategy. Having a setback temperature for unoccupied periods during the heating season or setup temperature during the cooling season will help to save energy. A pre-occupancy operation period will help to purge the building of contaminants that build up overnight from the outgassing of products and packaging materials. In hot, humid climates, care should be taken to avoid excessive relative humidity conditions during unoccupied periods.

4.3.4 Duct Insulation

All supply air ductwork should be insulated. All return air ductwork located above the ceiling and below the roof should be insulated. Any outdoor air ductwork should be insulated. All exhaust and relief air ductwork between the motor-operated damper and penetration of the building exterior should be insulated. Include a vapour barrier on the outside of the insulation where condensation is possible.

4.3.3 控制策略

控制策略的设计可以帮助四线节能。在已知建筑的哪个部分在哪个时间段的占用率较低的情况下，分时段计划十分有用。通风系统的控制可以与此策略绑定。在采暖季节无人使用的时间段设置较低的温度以及在制冷季节设置较高的温度都能实现节能。使用前运行时段将会帮助净化建筑物内由产品和包装材料一整夜所释放的污染物。在炎热潮湿的气候，应当注意避免建筑在无人使用时段的相对湿度过高。

4.3.4 风道绝缘

所有送风管道都应该绝缘。所有位于天花板之上、屋顶之下的回风管道都应该绝缘。所有户外空气管道都应该绝缘。所有电动风闸和建筑外部渗透之间的排气和减压管道都应该绝缘。如有冷凝的可能，在绝缘层外设置一层防潮层。

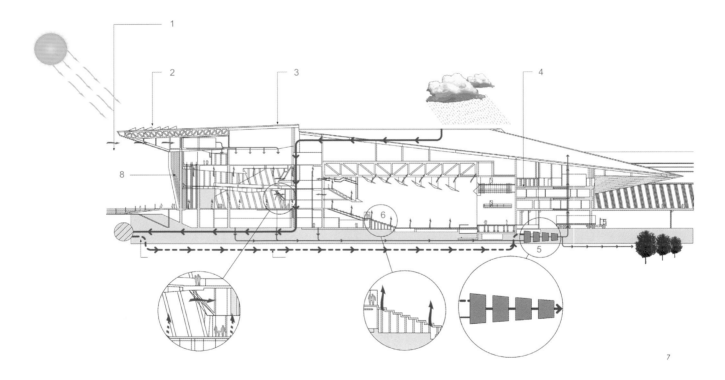

7. Diagram of energy saving and emission reduction

7. 节能减排示意图

1. Fresh Air Rates improve indoor air quality due to large volumes of fresh air
2. Solar Hot Water Solar panels provide 100% of public amenity hot water requirements
3. Sustainable Use of Building Materials, timber from renewable sustainable sources, materials and components have a high recycled content and minimal PVC utilisation
4. Low Volatile Organic Compounds (VOC) carpets, paints, adhesives and sealants to be low in VOC to enhance indoor air quality
5. Black Water Treatment Facility treats wastewater, rainwater and stormwater to Grade A quality for reuse in building, consequently reduces flow to sewer
6. Displacement Ventilation, low level air delivery and high level air exhaust provide excellent air change effectiveness and high indoor air quality at low energy consumption
7. Radiant Slab, Heating & Cooling Slab heated to provide energy efficient occupant thermal comfort and reduce air conditioning requirements
8. Expansive Glass Façade provides views and allows high degree of diffused natural light with spectrally selective glass

1. 大量的新鲜空气提升了室内空气质量
2. 太阳能热水器能满足100%的公共设施热水需求
3. 建筑材料的可持续应用，木材来自可再生森林，材料和构件具有高回收率，聚氯乙烯的使用最小化
4. 低挥发性地毯、涂料、胶合剂和密封剂，有助于提高室内空气质量
5. 黑水处理设施处理废水、雨水，使其达到A级可利用水平，从而减少污水排放
6. 置换通风，低层送风和高层排风在低能耗的前提下，实现了高效的空气置换率和良好的室内空气质量
7. 加热制冷辐射板的应用保证了热舒适度，同时减少了空调需求
8. 大面积的玻璃幕墙提供了良好的视野，并且以光谱选择玻璃提供了高等级的漫射自然光

4.3.5 Energy Recovery
Total energy recovery equipment can provide an energy-efficient means of dealing with the latent and sensible outdoor air cooling loads during peak summer conditions. It can also reduce the required heating of outdoor air in cold climates.

Exhaust air energy recovery can be provided through a separate energy recovery ventilator (ERV) that conditions the outdoor air before entering the air-conditioning or heat pump unit, an ERV that attaches to an air-conditioning or heat pump unit, or an air-conditioning or heat pump unit with the ERV built into it.

For maximum benefit, energy recovery designs should provide as close to balanced outdoor and exhaust airflows as is practical, taking into account the need for building pressurisation and any exhaust that cannot be incorporated into the system.

Exhaust for ERVs may be taken from spaces requiring exhaust (using a central exhaust duct system for each unit) or directly from the return airstream (as with a unitary accessory or integrated unit).

Where economizers are used with an ERV, the energy recovery system should be controlled in conjunction with the economizer and provide for the economizer function. Where energy recovery is used without an economizer, the energy recovery system should be controlled to prevent unwanted heat and an outdoor air bypass of the energy recovery equipment should be used. In cold climates, manufacturer's recommendations for frost control should be followed.

4.3.6 Exhaust Air
Central exhaust systems for restrooms, janitorial closets, etc., should be interlocked to operate with the air-conditioning or heat pump unit except during unoccupied periods. These exhaust systems should have a motorized damper that opens and closes with the operation of the fan. The damper should be located as close as possible to the duct penetration of the building envelope to minimise conductive heat transfer through the duct wall and avoid having to insulate the duct. During unoccupied periods, the damper should remain closed and the exhaust fan turned off, even while the air-conditioning or heat pump unit is operating, to maintain setback or setup temperatures.

4.3.7 Noise Control
Acoustical requirements may necessitate attenuation of the noise associated with the supply and/or return air, but the impact on fan energy consumption should also be considered and, if possible, compensated for in other duct or fan components. Acoustical concerns may be particularly critical in short, direct runs of ductwork between the fan and supply or return outlet. Where practical, avoid installation of the air-conditioning or heat pump units above areas that customers visit. Consider locations above less critical spaces such as storage areas, restrooms, corridors, etc.

4.4 Service Water Heating
The service water heating (SWH) equipment is considered to use the same type of fuel source used for the HVAC system.

4.4.1 Service Water Heating System Description
1. Gas-fired storage water heater. A water heater with a vertical or horizontal water storage tank. A thermostat controls the delivery of gas to the heater's burner. The heater requires a vent to exhaust the products of combustion.

4.3.5 能量回收
总能量回收设备可以在夏季峰值负荷时段以节能的方式处理户外空气显热和潜热负荷。能量回收还能在寒冷气候减少户外空气加热所需的能量。

排气能量回收可通过独立的能量回收通风设备（ERV）实现。该设备在户外空气进入空调或热泵机组之前对其进行调节。附属于空调或热泵机组的ERV以及空调系统或热泵机组内置的ERV也可以实现废气能量回收。

为实现利益最大化，能量回收设计应当尽量贴近平衡户外气流和排气气流。设计需要考虑到建筑的增压需求以及不能被该系统处理的排放物。

当节能装置配有ERV系统时，能量回收系统的控制应当与节能装置相结合，以实现节能装置的功能。当能量回收没有配备节能装置，能量回收系统的控制应当避免产生不必要的热量，同时应当使用能量回收设备的户外空气旁路。在寒冷气候，应当遵守制造商的推荐规范。

4.3.6 排气
除无人使用时段外，洗手间、工具房等空间的中央排气系统应当与空调或热泵机组实行连锁运转。排气系统应当配备与风扇运行进行对应开闭的电动挡板。挡板应尽量接近建筑外壳的管道渗透系统，以减少管壁的热传导、避免对管道进行绝缘。在无人使用的时段，即使空调或热泵机组仍在运转，挡板和排气扇也应保持关闭，以保持相对节能的温度。

4.3.7 噪音控制
声学要求促使设计师消减与送气、排气相关的噪音，但是风机能耗的影响也应在考虑之列。风机与送气口和回气口之间的短小、直接的管道噪音尤其明显。如果可行，尽量避免将空调或热泵机组安装在有顾客的区域。考虑将它们安装在仓库、洗手间、走廊等非重点区域。

4.4 生活用水加热
生活用水加热设备应与空调系统采用相同类型的燃料源。

4.4.1 生活用水加热系统概述
1. 储水式燃气热水器。此种热水器配有垂直或水平的储水箱，由恒温器控制燃气向热水器燃烧器的传送。热水器需要排放口来排放燃烧的产物。

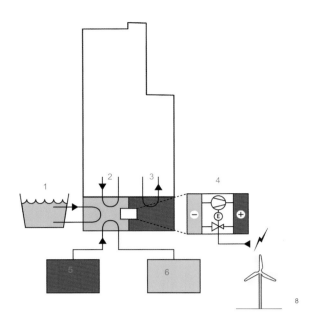

8. Diagram of water circulation

8. 水循环示意图

1. River Maas
2. Cold
3. Heat
4. Heat Pump
5. Heat storage
6. Cold storage

1. 玛斯河
2. 冷
3. 热
4. 热泵
5. 蓄热
6. 蓄冷

9. Diagram of water circulation

9. 水循环示意图

1. Cold maas water
2. Warm water from the warm reservoir
3. Cold water flows into the cold water well

1. 冷河水
2. 温水蓄水池中的温水
3. 冷水流入冷水井

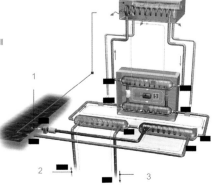

2. Gas-fired instantaneous water heater. A water heater with minimal water storage capacity. The heater requires a vent to exhaust the products of combustion. An electronic ignition is recommended to avoid the energy losses from a standing pilot.
3. Electric resistance storage water heater. A water heater consisting of a vertical or horizontal storage tank with one or more immersion heating elements. Thermostats controlling heating elements may be of the immersion or surface-mounted type. For typical retail applications, small water heaters are available from 2 to 20 gal.
4. Electric resistance instantaneous water heater. Compact, under-cabinet, or wall-mounted types with insulated enclosure and minimal water storage capacity; a thermostat controls the heating element, which may be of the immersion or surface-mounted type. Instantaneous, point-of-use water heaters should provide water at a constant temperature regardless of input water temperature.

4.4.2 Equipment Efficiencies
Efficiency levels are provided in the Guide for gas instantaneous, gas-fired storage, and electric resistance storage water heaters. For gas-fired instantaneous water heaters, the energy factor and thermal efficiency levels correspond to commonly available instantaneous water heaters.

The gas-fired storage water heater efficiency levels correspond to condensing storage water heaters. High-efficiency, condensing gas storage water heaters (energy factor > 0.90 or thermal efficiency > 0.90) are alternatives to the use of gas-fired instantaneous water heaters.

Electric storage water heater efficiency should be calculated as 0.99 − 0.0012 × water heater volume. Instantaneous electric water heaters are an acceptable alternative to high-efficiency storage water heaters. Electric instantaneous water heaters are more efficient than electric storage water heaters, and point-of-use versions will minimise piping losses. However, their impact on building peak electric demand can be significant and should be taken into account during design.

4.4.3 Pipe Insulation
All service water heating piping should be installed in accordance with accepted industry standards. Insulation should be protected from damage. Include a vapour retardant on the outside of the insulation.

10. Diagram of air circulation
10. 空气循环示意图

1. Flat solar collectors
2. Solar protection
3. Cross ventilation
4. Air convection
5. Solar radiation
6. Material radiation

1. 平板太阳能集热器
2. 遮阳
3. 交叉通风
4. 空气对流
5. 太阳辐射
6. 材料辐射

4.4.2 设备效率
本指南将阐述即热式燃气热水器、储水式燃气热水器和储水式电热水器的效率等级。即热式燃气热水器的能量因数和热效率等级相当于常用的普通即热式热水器。

储水式燃气热水器的效率等级相当于冷凝式储水热水器。高效的冷凝式燃气热水器（能量因数 > 0.90 或热效率 > 0.90）可以作为即热式燃气热水器的替代品。

储水式电热水器的效率计算公式为：0.99~0.0012 × 热水器容积。即热式电热水器可以作为高效储水热水器的替代品。即热式电热水器比储水式电热水器更加高效，而即用版本则可以将管道损失最小化。然而，它们的建筑的峰值电力需求的影响极大，在设计中应予以考虑。

4.4.3 管道绝缘
所有生活用水加热管道都应当采用公认的行业标准进行绝缘。绝缘应避免造成损害。绝缘的外层应添加水蒸气阻化剂。

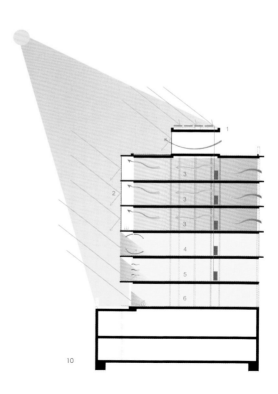

4.5 Others

In addition, "bonus savings" strategies to improve energy efficiency energy savings level are included for:

4.5.1 Exterior Façade Lighting

Exterior lighting should be turned off or reduced by at least 50% one hour after normal business hours in response to light pollution and light trespass concerns. Maintain lighting that is required for safety and security.

Limit exterior decorative façade lighting to 0.2 W/ft^2 of illuminated surface. This does not include lighting of walkways or entry areas of the building that may also light the building itself. Façade lighting can provide additional attention to the retailer and improve feelings of safety and security. Limit the lighting equipment mounting locations to the building and do not install floodlights onto nearby parking lot lighting standards. Use downward-facing accent and sign lighting to comply with light trespass and light pollution concerns.

4.5.2 Parking Lot Lighting

Parking lot lighting locations should be coordinated with landscape plantings so that tree growth does not block effective lighting from pole-mounted luminaires. Parking lot lighting should not be significantly brighter than lighting of the adjacent street.

For parking lot and grounds lighting, do not increase luminaire wattage in order to use fewer lights and poles. Increased contrast makes it harder to see at night beyond the immediate fixture location. Flood lights and non-cut-off wall-packs should not be used, as they cause hazardous glare and unwanted light encroachment on neighbouring properties. Limit lighting in parking and drive areas to not more than 360-watt pulse-start metal halide lamps at a maximum 25 ft mounting height in urban and suburban areas. Use cut-off luminaires that provide all light below the horizontal plane and help eliminate light trespass.

The use of cut-off luminaires and limiting overall site brightness also permits greater visibility of storefronts from more distant locations of the parking areas and adjacent roadways, permitting retailers greater opportunity to create off-site visual impact

4.5.3 Plug Loads

Building owners and other users can benefit from additional energy savings by outfitting retail stores with efficient appliances, office equipment, and other devices plugged into electric outlets. These "plug loads" can account for 4% to 40% of a retail building's annual energy requirements and energy expense. In addition to their own energy requirements, plug loads also are a source of internal heat gains that increase air-conditioning energy use. There is a large variation of these loads depending on the retail operation and the need for retail displays, communications equipment, computing requirements, product movement, and cleaning requirements.

Most retail operations have small computer networks that may include inventory systems and point-of-sale devices. In many retail operations, Internet terminals provide customers the ability to order products online. Retail stores also have offices for managers and staff; these spaces include devices such as fax machines, calculators, and copiers. Employee break rooms are often equipped with refrigerators, microwave ovens, and coffee makers. Many retail stores have vending machines that may or may not be publicly accessible.

4.5 其他

此外，以下是能够提升能效、节能等级的"附加节约"策略。

4.5.1 外墙照明

在营业时间结束后，应关闭外墙照明或至少将其调小到 50%，以避免造成光污染或光入侵。保留安全和保安照明。

将外墙装饰照明的照明表面平均功率限制在 0.2 W/ft^2。这不包括可能会点亮建筑的走道或入口区域明。外墙照明可以为销售商吸引更多的注意力，并且能提升安全感。将照明设备的位置限制在建筑上，不要将泛光灯安装在附近的停车场。利用向下的重点照明和标识照明来解决光入侵和光污染的问题。

4.5.2 停车场照明

停车场的照明位置应当与景观植被相协调，以保证树木生长不会阻挡灯柱的有效照明。停车场照明不应强于相邻街道的照明。

在停车场照明中，不要通过增加照明功率来减少灯柱数量。更强的对比让人眼更难看见近处固定位置之外的物体。不要使用泛光灯和非截光型壁灯，因为它们可能造成危险的眩光以及对周边设施的光入侵。将停车场和车道的照明限制在 360 瓦脉冲金属卤素灯，城市及郊区的灯具高度不应超过 25 英尺（约 7.62 米）。采用所有光线都低于水平面的截光型灯具来帮助消除光入侵。

截光型灯具的使用和整体场地亮度的限制还能增加店面从远处停车场和相邻街道的可见度，为销售商提供更多的场外视觉效应。

4.5.3 插座负荷

建筑业主和其他使用者可以通过为零售店铺配备高效的电器、办公设施和其他使用插座电源的设备实现额外的节能。这些"插座负荷"可占有建筑年度能源需求和能源开销的 4%～40%。除了它们自身的能源需求之外，插座负荷还是室内热增量的来源，能够增加空调能耗。根据经营方式和商品展示的必要、通讯设备、计算需求、商品移动以及清洁需求的不同，这些负荷有多个变种。

大多数经营方式都拥有小型计算机网络，包含库存系统和零售终端设备。在许多零售业务中，因特网终端为消费者提供了网上订货的选择。零售店还设有经理和员工的办公室，这些空间可能包含传真机、计算机、复印机等设备。员工休息室通常配有冰箱、微波炉和咖啡机。许多零售店还设有公开或仅限内部使用自动贩卖机。上蜡机、抛光机等清洁设备可以直接连接插

Cleaning equipment, such as waxers and buffers, can be directly connected or recharged. Telephone switches, fire and security alarm systems, and energy management systems all contribute to the building loads. Design teams should identify special loads in retail spaces (examples might include lighting displays in hardware stores and fish tanks in pet stores) and try to reduce these loads as appropriate. The team should take inventory in an existing similar facility to fully understand the loads and determine possible points for energy savings.

座或进行充电。电话交换机、火警和安全警报系统以及能源管理系统都会增加建筑负荷。设计团队应当明确零售空间的特殊负荷值（例如，五金店的照明展示、宠物店的鱼缸），并尽量减少这些负荷。设计团队应当制作现有类似设施的负荷清单，以全面了解建筑负荷问题并确定节能的可能性。

11. Diagram of energy saving and emission reduction
11. 节能减排示意图

1. Solar hot water and PV panels
2. Summer sun shaded & warming winter sun used for passive heating
3. Heating provided by occupancy and cooking in winter
4. Food and organic waster placed in bio-digester used as gas for cooking
5. Underfloor heating and cooling circuit powered by PV generation
6. Rainwater used in low flush WCs
7. Wind cowl provides passive wind assisted ventilation with coolness recovery
8. PV power is inverted to mains voltage to power lights and appliances
9. Solar thermal unit used to dry desiccant material for passive cooling
10. Exposed concrete in walls and ceilings for passive cooling
11. Air tightness line to ensure passive cooling/heating functions when windows are closed
12. Envelope surrounded by super insulation to keep warm in winter and cool in summer
13. Rainwater collected for irrigation
14. PV used to charge batteries which can change from the site wide grid or export to it depending in energy generation

1. 太阳能热水器和光电伏板
2. 夏季遮阳&冬季被动式太阳能采暖
3. 冬季，使用者和烹饪所产生的热量
4. 食品和有机废弃物被投入生态处理池，产生的沼气用于烹饪
5. 地面采暖和制冷循环系统，由光伏发电供能
6. 雨水被用于低流量马桶
7. 通风罩提供辅助通气和冷气回收
8. 光伏发电被转化为电源电压，为电灯和电器供电
9. 将太阳能热机组用于干燥材料上进行被动制冷
10. 墙面和天花板的清水混凝土可实现被动制冷
11. 空气密度线保证了关窗时被动制冷和采暖功能的实现
12. 建筑外壳的隔热层保证室内冬暖夏凉
13. 雨水收集，用于灌溉
14. 光伏发电被转化到充电电池里，可以应用在建筑场地之内，也可输出到电网中

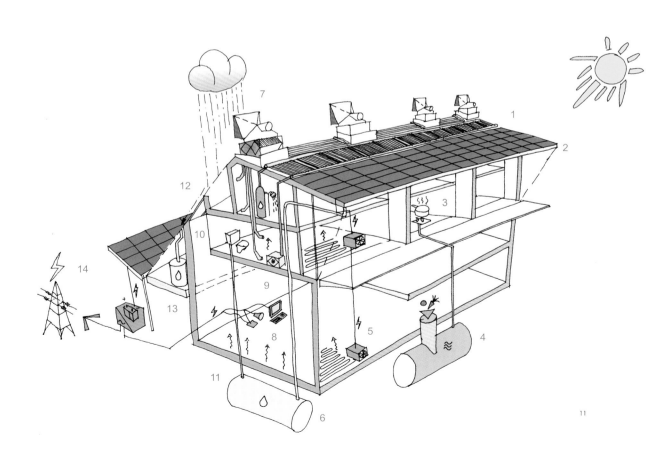

Completion date: 2011
Location: Graz, Austria
Designer: LOVE architecture
Photographer: Jasmin Schuller
Construction area: 1,500sqm

Climate Protection Supermarket
气候保护超市

Key words: A Successful Example of Energy Self-sufficiency
Increasing attentions are paid to harmonious environment and sustainable development. The project is Austria's first energy self-sufficient supermarket. The supermarket uses a hydropower turbine for energy generation, which could create more energy than the store uses, and the excess energy can be fed back into the grid. The store has won a gold certification from the OGNI (Austrian Green Building Council). Besides using dismountable and reusable building materials, room ventilation, heat recovery, a sectional foundation slab for cooling and heating, Led lighting and daylight control and other traditional low-carbon technologies, the designer also design the roof as a fifth façade according to local climate to maximise green space and achieve a hydrologic balance by leaching all of the surface water on the property. In addition, the delivery area is enclosed to minimise noise emissions.

主题词：能源上自给自足的成功范例
和谐的环境和可持续发展越来越被重视，本项目是首个能源自给商店，水力涡轮机发电，为商店创造出盈余的能量，多余的能量可以被输送入电网，取得了OGNI（奥地利绿色建筑委员会）的黄金认证。除了延续使用可分离和可重复使用的建筑材料；实现室内通风、热回收以及用于制冷和供暖的分段基础板；使用LED技术和日光控制系统等传统低碳环保手段外，设计师还根据当地的气候将屋顶设计成第五个立面，建造尽可能多的绿色空间，用以过滤建筑表面的水，实现水分平衡。此外，装卸区实现了封闭，减少了噪音排放。

Design Points:

Organic products and a healthy lifestyle are mega trends in our society and therefore in the food industry as well. When building supermarkets, environmental friendliness and sustainability are becoming increasingly important. Economically and ecologically sustainable construction and operation minimize the ecological footprint and reduce the life-cycle cost of buildings.

This market is a 3rd generation climate protection store with a gold certification from the ÖGNI (Austrian Green Building Council). In fact, the site produces more energy than the store uses, making it Austria's first energy self-sufficient supermarket.

Architecturally speaking, the structure consists of a simple folded shell that arches over the triangu-

lar-shaped property.

The store opens onto the parking lot in all three dimensions, to the front, sideways and upwards, which creates very broad and inviting entry from this direction.

The building envelope itself features slight folds and wrinkles. This creates a different effect from each perspective, and the building thereby achieves a significant dynamic and tension: "like an athlete before jumping."

For the façade, the goal was to create a strong haptic quality with materials that convey the themes of climate protection and naturalness, but also modernity and innovation. The façade consists of galvanised sheet steel and wood. Due to their contrast, the two materials convey the different themes, while also creating additional excitement.

1. North exterior view
2. Back door
3. South exterior view

1. 建筑北侧外观
2. 后门
3. 建筑南侧外观

1. West exterior view
2. Simple folded shell

1. 建筑西侧外观
2. 简洁的折叠形外壳

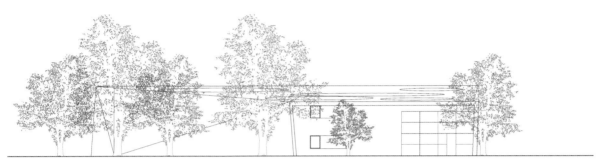

Southwest elevation　西南立面图

Northeast elevation　东北立面图

Northwest elevation　西北立面图

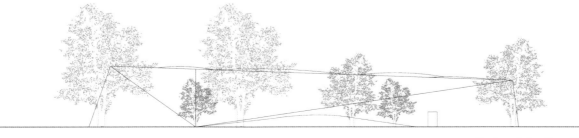

Southeast elevation　东南立面图

The roof was designed as a fifth façade, with circular, hill-shaped green spaces.

Climate protection store – technical implementation:
In order to achieve the goal of a climate protection store, a wide variety of measures were necessary, which can be summarised in five large areas:
The highly-insulating building envelope minimises both energy loss and energy input.
The building technology includes room ventilation, heat recovery and a sectional foundation slab for cooling and heating; lighting with LED technology and daylight control systems.
The use of sustainable, separable and reusable building materials, preferably solvent-free and non-toxic.

The micro climate at the site
This means the creation of as many green spaces as possible, including on the roof, which were designed to achieve a hydrologic balance by leaching all of the surface water on the property. In addition, the delivery area is enclosed to minimise noise emissions.

The energy generation
In addition, a photovoltaic power plant is located in the parking lot, and a hydropower turbine will be installed for energy generation. These features create more energy than the store uses, and the excess energy can be fed back into the grid.

The overall result is an energy self-sufficient climate protection store that offers an inviting atmosphere for shoppers and a high-quality workplace for employees.

有机产品和健康的生活方式是目前社会的主要流行趋势，因此在食品工业领域亦是如此。在超市设计中，环保意识及可持续发展理念正变得越来越重要。经济的和生态可持续的建设运营模式可以减小生态足迹和降低建筑物的寿命周期成本。

这个市场是一个第三代气候保护店铺，得到了OGNI的黄金认证（奥地利绿色建筑委员会）。事实上，店铺产生的能量多于自身消耗，因此，它成为奥地利第一个能源自给自足的超市。
从建筑上讲，其结构是由一个简单的折叠外壳，性能类似拱起的三角形。

商店有三个方向通往停车场，无论是往前、往侧及往上都可以，因此进入店铺的空间宽阔且方便。
建筑围护结构本身具有轻微的折痕和褶皱。从每一个角度看都会有一个不同的视觉效果，从而得到了显著的动感和张力："像运动员的赛前热身。"

至于店铺外观，设计师的目标是通过材料的使用来传达保护气候和自然的主题，建立一种强烈的触觉质感，同时兼顾设计的现代和创新。外表皮由镀锌钢板和木材组成。由于它们的强烈对比感，这两种材料传达出不同的主题，同时也带来了额外的刺激。

屋顶作为第五个面，被设计成圆形、山状的绿色空间。

气候保护商店 – 技术实施：
为了实现建成气候保护商店这一目标，采取各种各样的方法是必要的，基本上可以归纳为五大方面：
高绝缘的建筑外围结构最大限度地减少能量损失和能量输入。
使用的技术包括室内通风，热回收和一个截面基础板供冷却和加热；照明采用LED技术和日光控制系统。

利用可持续、可分离和可重复使用的建材,并优先选择无溶剂的和非毒性的材料。

当地的微气候

这意味着要创造包括屋顶在内的尽可能多的环保空间,充分利用所有浸出的地表水,旨在实现水文平衡。此外,送货区域是封闭的,以减少噪音排放。

能量产生

此外,光伏电站位于停车场,并安装了用于发电的水电涡轮机。这些特性创造出的能量比店铺的消耗更多,多余的能量可以回到输电网。

这个能源自给自足的气候保护商店最终顺利建成了,它为购物者提供了一个温馨的氛围,也为员工提供了高品质的工作环境。

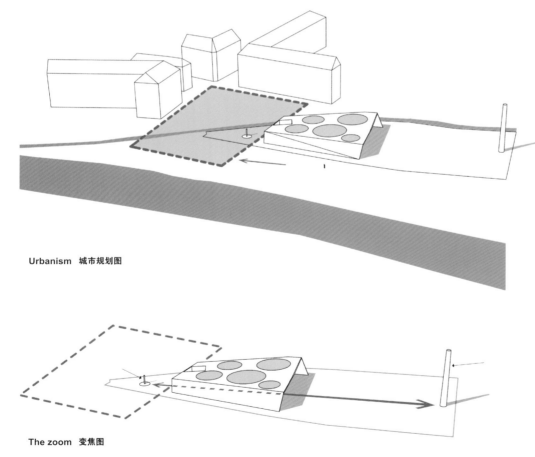

Urbanism 城市规划图

The zoom 变焦图

1. The shopfront
2. Interior view

1. 店面
2. 室内设计

1. Entrance
2. Vegetables and fruits
3. Sales counter for meat, sausages, cheese and pastries
4. Chiller cabinet
5. Freezers
6. Checkout counters
7. Delivery
8. Storage
9. Women's bathroom
10. Men's bathroom
11. Men's changing room
12. Office
13. Recreation room
14. Women's changing room
15. Preparation room for pastries
16. Cooling chamber for pastries
17. Cooling chamber for meat
18. Preparation room for meat
19. Utility room

Site plan　总平面图

1. 入口
2. 果蔬区
3. 肉类、香肠、奶酪、面点区
4. 冰柜
5. 冰箱
6. 收银台
7. 交货区
8. 仓库
9. 女洗手间
10. 男洗手间
11. 男更衣室
12. 办公区
13. 娱乐室
14. 女更衣室
15. 面点准备室
16. 面点冷冻室
17. 肉类冷冻室
18. 肉类准备室
19. 杂物间

Completion date: **2011**	Designer: **Comas pont arquitectes slp.**	Site area: **1,689.19sqm**
Location: **Manlleu, Spain**	Photographer: **Jordi Comas**	Construction area: **2,187sqm**

Municipal Market 市政市场

Key words: Low-carbon Green in Multiple Aspects
The project completed all of its construction in 8 months, achieving maximum saving of time and money. The folded form allows natural ventilation and daylighting. The project applies advanced construction system: façades, roof and interior surface uses natural materials to meet the structure requirements and they are easy to dismantle and recycle. Roof water collection, renewable energy and geothermal energy heating realize the project's low-carbon goal in multiple aspects.

主题词：多方位的实现低碳环保
该项目在8个月内完成了所有的施工，最大化的节约了时间和资金。建筑的形态错落有致，正是这样高低起伏的设计使自然通风和自然光照明得以充分实现。项目应用了先进的施工系统；立面、屋顶及室内外装材料选择了天然的环保材料，满足了结构的要求，并可拆卸和回收；实现了屋顶的水收集系统、可再生能源和地面热能采暖，在各个环节和角度实现了低碳的初衷。

Design Points:

The new fresh products market is a Mediterranean building: the sales area is a large open space with the light as a protagonist, a controlled light filtered through a wooden slats system of the façades and inner skin panels and slats of wood shavings.

A continued zinc skin unifies roof and facades and it folds generating different highs that allow light entry and the all building ventilation. This skin folds itself creating porches to the main entrances. Fragmented cover dialogues in height with environment's buildings and hide its big scale.

The different heihgts respond to specific programme necessities housed inside: the big sales space, opened through prefabricated metal truss arches, has the whole high and lineal windows on cover folds.

Logistic zone is situated on the west side and it has two floors with lower high, as same as the east strip where activities which worked independently from the market are situated, and therefore they can be a sector: restaurant, bar, polyvalent rooms and spaces to municipal associations.

This building is a key item to rehabilitate the neighbourhood. The urbanisation respects streets rank and emphasises Pintor Guardia street like a central neighbourhood axis, which is a pedestrian street now and connects equipments and public spaces.

The bottom floor is almost transparent and emphasises visual connection between interior and exterior. This is reinforced with continuity on the interior pavement to the exterior below porches. The cover strips are projected on the pavement with granite black lines which continue at exterior, therefore connectivity is emphasised and it is easy to define back park zone, tree areas, the hard urbanisations surface and all exterior furniture situation.

The designers used a dry semi-industrialised construction system to fulfill short timing (8 months) and minimise the waste. The construction system includes metal structure, sandwich wood panels in big format, wood slats, zinc roof and facades and interior OSB panels. All materials have natural finishes (wood and zinc). Passive systems (re-

1. Main façade
2. East corner
3. Access

1. 主立面
2. 建筑东角
3. 入口

newable energy heating and cooling systems, geothermal energy, heat recover machines), roof water collect tank and the possibility of dismantle and recycle all building materials let to catch with an excellent energetic building.

1. Urbanisation   1. 城市环境
2. Roof detail   2. 屋顶细部设计
3. Roof   3. 屋顶
4. West corner   4. 建筑西角

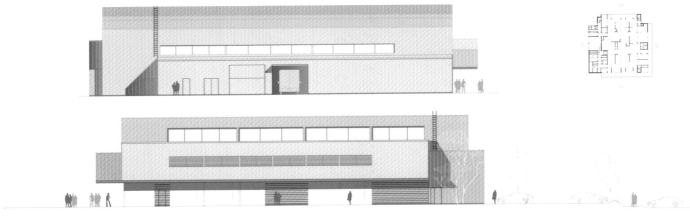

Elevations 立面图

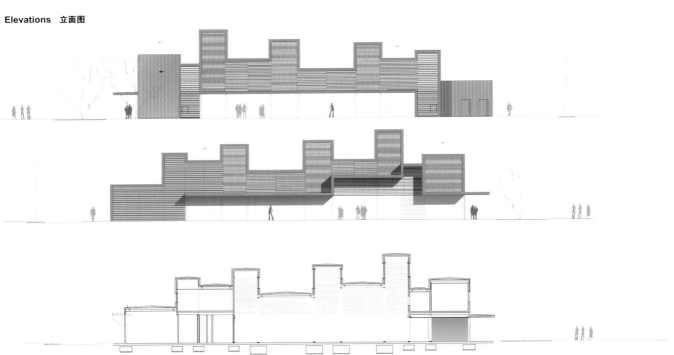

Section 剖面图

新建的生鲜市场为典型的地中海建筑：大型开放式售货区以光线为特色，光线透过外立面的木条和内墙的壁板及刨花板渗透进来。

连续的锌板将屋顶和外立面统一成一个整体，锌板的折叠起伏让阳光进入室内，并保证了建筑的通风。这层建筑外壳通过折叠形成了主入口的门廊。分散起伏的屋顶与周边的建筑高低呼应，也掩藏了建筑的巨大规模。

不同的高度呼应着建筑内部不同的功能区：大面积的售货空间通过预制金属桁架拱顶开放，两侧拥有垂直的落地玻璃。

物流区位于西侧，其所在的两层楼高较低，与东侧区域相同。东侧区域用于与市场无关的活动，如餐厅、酒吧、多功能厅和市政协会空间等。

建筑是复兴周边社区的重要项目之一。城市化进程以 Pintor Guardia 为中轴，将其划为步行街，起到了连接重要设施和公共空间的作用。

建筑低层基本是透明的，强调室内外的视觉连接。门廊下连接室内外的走道进一步突出了二者的连续性。屋顶的陶文投射到走道的黑色花岗岩分割线上，延续了室内外的协调性。这种设计也突出了后方的公园区、树木、市政地面铺设和所有室外休闲设施。

设计师采用半工业化建造系统来实现短期施工，同时减少了建筑废料的产生。建造系统包括金属结构、大型夹心模板、木板条、锌屋顶和内外墙的刨花板。所有材料都有天然饰面（木纹和金属纹）。被动系统（可再生能源加热和制冷系统、地热能源、热恢复机器等）、屋顶雨水收集系统以及对建筑材料的回收都使得项目成为了一座优秀的节能建筑。

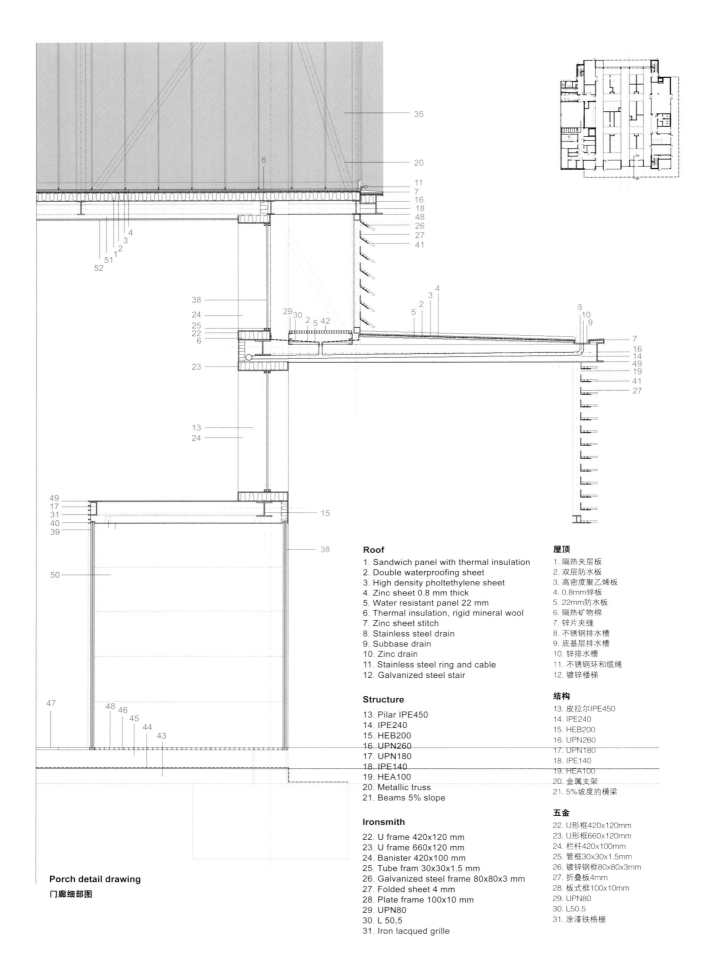

Roof
1. Sandwich panel with thermal insulation
2. Double waterproofing sheet
3. High density pholtethylene sheet
4. Zinc sheet 0.8 mm thick
5. Water resistant panel 22 mm
6. Thermal insulation, rigid mineral wool
7. Zinc sheet stitch
8. Stainless steel drain
9. Subbase drain
10. Zinc drain
11. Stainless steel ring and cable
12. Galvanized steel stair

Structure
13. Pilar IPE450
14. IPE240
15. HEB200
16. UPN260
17. UPN180
18. IPE140
19. HEA100
20. Metallic truss
21. Beams 5% slope

Ironsmith
22. U frame 420x120 mm
23. U frame 660x120 mm
24. Banister 420x100 mm
25. Tube fram 30x30x1.5 mm
26. Galvanized steel frame 80x80x3 mm
27. Folded sheet 4 mm
28. Plate frame 100x10 mm
29. UPN80
30. L 50,5
31. Iron lacqued grille

屋顶
1. 隔热夹层板
2. 双层防水板
3. 高密度聚乙烯板
4. 0.8mm锌板
5. 22mm防水板
6. 隔热矿物棉
7. 锌片夹缝
8. 不锈钢排水槽
9. 底基层排水槽
10. 锌排水槽
11. 不锈钢环和缆绳
12. 镀锌楼梯

结构
13. 皮拉尔IPE450
14. IPE240
15. HEB200
16. UPN260
17. UPN180
18. IPE140
19. HEA100
20. 金属支架
21. 5%坡度的横梁

五金
22. U形框420x120mm
23. U形框660x120mm
24. 栏杆420x100mm
25. 管框30x30x1.5mm
26. 镀锌钢框80x80x3mm
27. 折叠板4mm
28. 板式框100x10mm
29. UPN80
30. L50.5
31. 涂漆铁格栅

Porch detail drawing
门廊细部图

Skylight detail drawing
天窗细部图

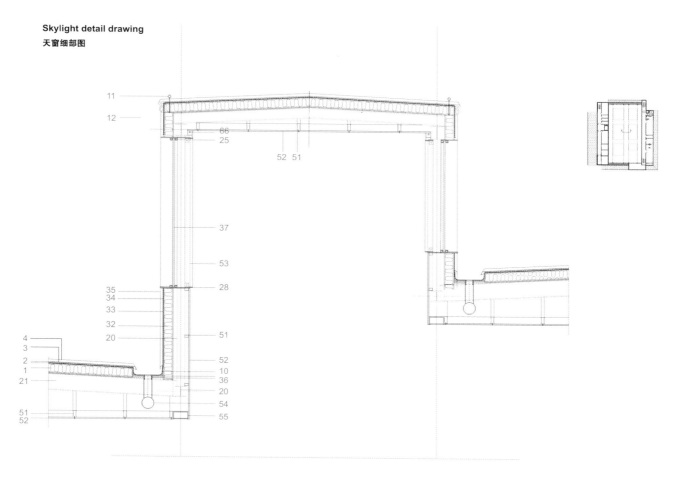

Closing
32. Waterproof sandwich panel with thermat insulation
33. Double waterproof sheet
34. Breathable sheet, vapor barrier
35. Zinc sheet
36. Thermal insulation, rigid mineral woof
37. Climalit glass with white butyral
38. Climalit glass with transparent butyral
39. Laminated glass
40. Stainless steel U profile
41. Larch slats 190x27 mm
42. Galvanized steel grate pavement

Cladding
43. Subbasement
44. Waterproof textile sheet
45. Reinforced concrete girder 20 cm thick
46. Layer of mortar
47. terrazzo pavement 40x40 cm
48. mat pavement
49. Lacqued iron sheet ceiling
50. Lacqued iron sheet
51. Galvanized steel profile
52. OSB panel 15 mm thick
53. OSB slats 10 cm wide and 15 mm thick

Installations
54. PVC drain pipe
55. Flourescent lighting mounted on rail

密封
32. 防水夹层，配隔热层
33. 双层防水板
34. 透气板，隔水蒸气
35. 锌板
36. 隔热层，矿物棉
37. 防风玻璃，配白缩丁醛涂层
38. 防风玻璃，配透明缩丁醛涂层
39. 夹层玻璃
40. U形不锈钢
41. 松木条190x27mm
42. 镀锌钢条铺面

包层
43. 底基层
44. 防水纺织薄板
45. 20cm厚钢筋混凝土梁
46. 砂浆层
47. 水磨石铺面40x40cm
48. 衬垫铺面
49. 喷漆铁皮天花板
50. 涂漆铁板
51. 镀锌钢
52. 15mm厚OSB板
53. 10cm宽，15mm厚OSB板

安装设施
54. 聚氯乙烯排水管
55. 栏杆上的荧光灯

1. Slats detail   1. 木板条细部设计
2. Ceiling   2. 天花板

Index / 索引

Castel Veciana Arquitectura
Website: www.castelveciana.com
Phone: +34 933 568818
Fax: +34 933 568817

Comas pont arquitectes slp.
Website: www.comas-pont.com
Phone: +34 93 889 04 67
Fax: +34 93 889 04 67

IOSA GHINI ASSOCIATES
Website: www.iosaghini.it
Phone: +39-051 236563
Fax: +39-051 237712

Jean Verville
Website: www.jeanverville.com
Phone: +351 226 163 408
Fax: +351 226 163 408

Jean Verville architecte
Website: www.jeanverville.com
Phone: 514 622 5722

KEY OPERATION INC. / ARCHITECTS
Website: http://www.keyoperation.com
Phone: 03.5724.0061
Fax: 03.5724.0062

klab architect
Website: www.klab.gr
Phone: +30 - 210 3211139
Fax: +30 - 210 3211155

kleboth lindinger dolling
Website: www.kld.as
Phone: +43 732 77 55 84
Fax: +43 732 77 55 84 88

Klein Dytham architecture
Website: http://klein-dytham.com
Phone: +81-3-5795-2277
Fax: +81-3-5795-2276

LOVE architecture and urbanism ZT GmbH
Website: www.love-home.com

MARZORATI RONCHETTI
Website: www.marzoratironchetti.it
Phone: +86 031 714147
Fax: +86 031 705060

monovolume architecture + design
Website: www.monovolume.cc
Phone: +39 0471 050 226
Fax: +39 0471 050 227

nendo
Website: www.nendo.jp
Phone: +81-(0)3-6661-3750
Fax: +81-(0)3-6661-3751

Lukas Göbl - Office for Explicit Architecture
Website: www.explicit-architecture.com
Phone: +43 (0)1 2764418

plajer & franz studio gbr
Website: www.plajer-franz.de
Phone: +89 (0)30 6165580
Fax: +89 (0)30 61655819

PROCESS5 DESIGN
Website: http://process5.com/
Phone: +81-6-6536-6336
Fax: +81-6-6536-0660

RTKL Associates Inc.
Website: www.rtkl.com

Sako
Website: www.sako.co.jp
Phone: +81-3-6265-1812

Standard
Website: www.standard-la.com
Phone: 323 662 1000
Fax: 323 662 0199

Studio Kalamar d.o.o.
Website: www.kalamar.si
Phone: +386 1 2410 470
Fax: +386 1 2410 473

x architekten
Website: www.xarchitekten.com
Phone: +43(0) 732 791607
Fax: +43(0) 732 7916075

Van den Pauwert Architecten BNA
Website: www.pauwert.nl
Phone: 0402812782

WHITESPACE
Whitespace Ltd
Phone: +66 2 235 2500
Fax: +66 2 235 2504

Architect Zarir Mullan & Seema Puri - Mullan
Website: www.seza.in
Phone: +91-22-24474780
Fax: +91-22-24474784